The Music Mixing Workbook

Exercises To Help You Learn How to Mix On Any DAW

A Companion To The Mixing Engineer's Handbook

By Bobby Owsinski

BOMG Publishing

The Music Mixing Workbook
Exercises To Help You Learn How to Mix On Any DAW
by Bobby Owsinski

Published by:
Bobby Owsinski Media Group
4109 West Burbank, Blvd.
Burbank, CA 91505

© Bobby Owsinski 2021
ISBN 13: 978-1-946837-11-0

Please note that much of this publication is based on personal experience and anecdotal evidence. Although the author and publisher have made every reasonable attempt to achieve complete accuracy of the content in this Guide, they assume no responsibility for errors or omissions. Also, you should use this information as you see fit, and at your own risk. Your particular situation may not be exactly suited to the examples illustrated herein; in fact, it's likely that they won't be the same, and you should adjust your use of the information and recommendations accordingly.

Any trademarks, service marks, product names or named features are assumed to be the property of their respective owners, and are used only for reference. There is no implied endorsement if we use one of these terms.

Finally, nothing in this book is intended to replace common sense, legal, medical or other professional advice, and is meant to inform and entertain the reader.

Books for educational use can be ordered from Ingram or Kortext, or email office@bobbyowsinski.com.

Table Of Contents

Title 1

Legal 3

Table Of Contents 5

Introduction 11

    FREE EXERCISE TRACKS.....................................12

1. Mixing Basics 13

A Brief History..13

Basic Signal Flow 15

    Typical Analog Console Signal Flow......................15

    Typical DAW Mixer Signal Flow 17

2. Monitoring 19

The Listening Environment 19

    Determining The Listening Position 19

    Acoustic Quick Fixes 20

Basic Monitor Setup 23

How To Listen...28

    Basic Listening Technique 28

    How Loud (Or Soft) Should I Listen?...................30

    Listening On Several Speaker Systems 32

    Listening In Mono 33

3. Session Setup 35

Prepping Your Mix 35

    Make A Session File Copy 35

    Arrange Your Tracks 36

    Insert Section Markers 37

    Cleaning Your Tracks 38

    Fix The Timing 39

    Set Up Groups Or Subgroups............................39

    Set Up Your Effects 40

    Insert Compressors And Limiters 41

Prepping Yourself 44

    Play Something You Know 44

    Take Notes 44

    Make Yourself Comfortable................................45

4. Mixing Mechanics 47

The Overall Mixing Approach 47
Developing The Groove 47
Emphasizing The Most Important Elements 48
Putting The Performers In An Environment 49

The 6 Elements Of A Mix 51
Additional Ear Training Exercises 56

The Parts Of A Musical Arrangement 56

5. Balance 61

Building The Mix 61
The Mix By Ear Method 61
The Mix By Meters Method 62
The Drums 64
The Percussion Element 72
The Bass Element 74
The Vocals 76
Keyboards 79
Guitars 80
Horns 82
Loops And Samples 83

Mixing By Muting 84

6. Panning 87

The Three Main Panning Areas 87
Big Mono 88
Low Frequencies In The Center 89
Panning Stereo Instruments 90
Panning The Drums 92
Panning Percussion 95
Panning The Bass 96
Panning Guitars 96
Panning Keyboards 98
Panning Vocals 99

7. Dynamics Processors 103

Compression 103
Compression Basics 103
Compressor Operation 107

Limiting 109

Compressing The Various Instruments 110
    Compressing The Drums *110*

Parallel Compression 119
    Compressing The Bass *120*

    Compressing Guitars *123*

    Compressing Keyboards *126*

    Compressing Vocals *127*

    Compressing Other Mix Elements *129*

    Compressing Loops And Samples *129*

De-essers 130
Gates 131

8. Using The EQ **133**

Equalization Basics 133
    EQ Parameters *134*

    A Description Of The Audio Bands *135*

    The Magic High-Pass Filter *138*

Using The Equalizer 141
    Subtractive Equalization *141*

    Juggling Frequencies *142*

EQing Various Instruments 145
    Equalizing The Drums *145*

    Equalizing The Bass Mix Element *146*

    Equalizing The Vocal *147*

    Equalizing The Electric Guitar *151*

    Equalizing The Acoustic Guitar *153*

    Equalizing The Piano *155*

    Equalizing The Organ *156*

    Equalizing Strings *157*

    Equalizing Horns *157*

    Equalizing Percussion *158*

The Principles of Equalization 159

9. Adding Reverb **161**

Reverb Basics 161
    Typical Reverb Parameters *162*

    Timing A Reverb To The Track *165*

Reverb Setup 168
    The Two Reverb Quick Setup Method *168*

Adding Reverb To Instruments 170
   Adding Reverb To The Drum Kit 170
   Adding Reverb To The Bass 176
   Adding Reverb To The Vocal 179
   Adding Reverb To Guitars 180
   Adding Reverb To Keyboards 181
   Adding Reverb To Strings 183
   Adding Reverb To Horns 184
   Adding Reverb To Percussion 184
Layering The Mix 185

10. Adding Delay 187
Delay Basics 187
   Typical Delay Parameters 187
   The Haas Effect 189
   Timing The Delay To The Tempo Of The Song 189
Delay Setup 195
   The Three Delay Full Setup Method 195
Adding Delay To Other Mix Elements 196
   Adding Delay To The Vocals 196
   Adding Delay To The Guitar 198
   Adding Delay To Keyboards 199
   Adding Delay To The Drum Kit 199
   Adding Delay To Other Mix Elements 202

11. Modulation Effects 203
Modulation Basics 203
   Types Of Modulation 203
   Parameter Settings 206
   Modulation Setup 207
Modulation On Instruments 207
   Modulation On Guitars 207
   Modulation On Keyboards 209
   Modulation On Vocals 211
   Modulation On Strings 214
   Modulation On Other Mix Elements 215

12. Creating Interest 219
Developing The Groove 219
   Finding The Groove 219

Establishing The Groove...220

Emphasizing The Most Important Element 221
Finding The Most Important Element...221

Emphasizing The Most Important Element.................................222

Making A Mix Element Interesting..223

Using Saturation 226

13. The Master Mix...229

Mixing With Subgroups 229
Individual Fader Levels...231

The Master Level Meters 231
Types Of Meters 231

Mix Buss Levels...233

Mix Buss Processing 234
Mix Buss Compressor Settings..234

Mix Buss Limiting...237

Stay Away From Hypercompression!..239

How Long Should My Mix Take?..240
How To Know When Your Mix Is Finished 241

14. Building The Mix In Order 243

Visualize Your Mix 243
The 10 Steps To Creating A Mix..245
Step 1: Prep Your Tracks 245

Step 2: Insert Master Buss Processing......................................245

Step 3: Insert The High Pass Filter...245

Steps 4 and 5: Set levels And Set Compression 245

Step 6: Tweak The Master Buss 245

Step 7: Begin EQing 246

Step 8: Add Effects 246

Step 9: Automate 246

Step 10: Tweak The Buss Processors 246

On Your Mixing Journey 246
About Bobby Owsinski 247
Other Music-Related Books By Bobby Owsinski....................248

Bobby Owsinski Lynda.com Video Courses 249

Bobby Owsinski Online Coaching Courses 250

Bobby Owsinski's Social Media Connections 251

Download Your Free Exercise Tracks..252

Introduction

Welcome to the Music Mixing Workbook.

If you're reading this book, it's probably because you're either new to mixing and aren't sure what to do, or your mixes aren't anywhere near where you'd like them to be. I can help, but only if you're willing to put the time in and do the work, because this book is built mostly around mixing exercises.

Mixing is one of those things that you can't learn by reading, you have to learn by doing. Like almost anything else, the more you do it, the better you become at it. The problem is that many new mixers just starting out don't know where to begin, and those that already know a little don't know what to do to get better.

That's were I come in.

I'll take you through all the areas of mixing that will not only get you going on the right track, but cut a lot of time off your learning curve as well.

Rather than just covering the theory of mixing, I'll have you pushing virtual faders and moving virtual controls so that you'll learn the secrets of mixing faster and easier than you ever thought possible (the exercises are designed to work on any DAW, but will also work on a traditional analog or digital console too).

Along the way I'll tell you the reasons behind the exercises as well as the specifics of what you're learning, but almost every topic in the book is strictly hands-on.

You still have to put in the work, but if you're making your music sound better, you'll have a lot of fun along the way too.

Hopefully you'll find the exercises enjoyable and a great learning experience. Sometimes an exercise may seem a bit off the wall at first, but be assured that it's there for you to learn why that action might or might not get you the results that you're hoping for.

My goal is to teach you all the things that make a mix sound great, and *that includes everything you can do to make it sound bad as well.* You may find you'll learn more from the later than the former.

Also note that this book won't cover basic theory or signal flow of a mixing console. For that you need to refer to the instruction manual of your DAW, or a book like my *The Mixing Engineer's Handbook* for more insight into this area. If fact, *The Mixing Engineer's Handbook* is also a great place to continue learning after you've finished this book, as it will give you a number of additional techniques that you can experiment with.

If you already have *The Mixing Engineer's Handbook*, you'll find that this Workbook is meant to be a companion to further illustrate the concepts found in that book. If not, you'll find that it has more in-depth explanations, additional material, and interviews with top mixers that you'll find both informative and enjoyable.

So get your DAW warmed up as we dig in deep into the world of mixing.

FREE EXERCISE TRACKS

Download the sample tracks that go along with the exercises in this book at:

bobbyowsinskicourses.com/sessions

There you'll find 2 songs with a variety of mix elements to work on in both Pro Tools and Logic sessions. You can also download the raw tracks into any other DAW of your choice.

Chapter 1
Mixing Basics

There are many great mixers that can't tell you exactly what they do because they do it all entirely by feel, intuition and experience. These qualities are usually developed over a long period of time with a lot of experimentation thrown in. You want to get there faster than that, which is why you're reading this book, so a more structured method of learning about mixing is required. Before you start moving faders, it's important to understand why you're doing it, and what you're listening for first.

A Brief History

Before we cover the mechanics of mixing, it's important to understand how this art has developed over the years.

In the early days of recording in the 50's, mixing certainly was far from how we know it today, since the recording medium was mono and a big recording date used only 4 microphones. With the development of tape machines with more and more tracks, larger mixing consoles with more and more channels were required as well. Soon consoles were so large that they needed computer automation and recall in order to adequately manage these larger track and channel counts. What we consider "mixing" evolved from a simple balancing of a few microphones fed into one or two tracks with very few outboard tools available, into a highly creative process managing more than a hundred tracks and channels with racks and racks of outboard gear. This required that the mixer develop not only a new approach to mixing, but a whole new way of listening as well.

Once tape machines were capable of more audio tracks, two major transformations occurred.

- Recording changed from capturing a performance of many musicians and vocalists in the studio, to creatively molding a song through overdubs. For instance, most of the hit songs of the 1950's and early 60's (like anything from The Beatles, RCA Records in Nashville, Motown Records, or producer Phil Spector) were recorded with virtually all the musicians playing in the studio at the same time. The overdubs consisted of vocals (lead and background at the same time), maybe horns or strings, and occasionally another lead instrument. From the 1970's onward when the 16 track, and later 24 track, tape machines came into widespread use (see Figure 1.1), most songs were built from the ground up by first recording the rhythm section of bass and drums (sometimes only the drums), and then adding everything else through a series of overdubs. This practice continues today.

Figure 1.1: A Studer A820 24 Track Tape Machine

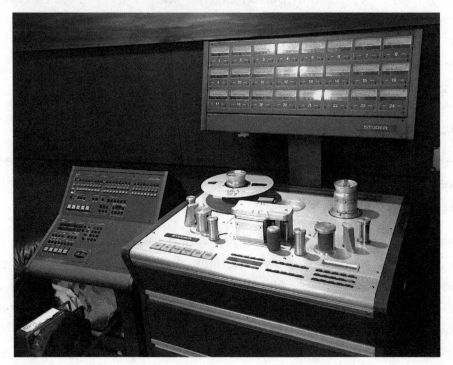

- The emphasis shifted from the bass being the foundation of the song to the drums. Up until the end of the 60's, the bass was considered the most important instrument in the mix and everything was built around it. Beginning with the later Beatle and Pink Floyd albums, the drums came to the forefront, thanks to individual miking (the drums where miked with only two mics previously). Today, more attention is given to the drum sound than ever before when they're a part of a song.

> **Exercise 1.1**: *Identifying Different Recording Techniques and Eras*

A) Play any song from the 1950's or early 60's.

☐ Notice how everything seems small and the instruments aren't that distinct?

☐ Notice how thin the sound is?

B) Now play any song from the 70's to today.

☐ Notice how much bigger it sounds?

☐ Notice how you can hear every instrument more distinctly?

C) Play any song from the 1950's or early 60's.

☐ Notice how the bass is at the forefront of the mix?

☐ Listen to where the drums are in relation to the other instruments in the mix.

D) Now play any song from the 70's to today.

☐ Notice where the drums are in the mix.

Basic Signal Flow

Whether you're using a digital audio workstation application (the name is usually just shortened to DAW) or a traditional hardware console, it's important to understand the way the signal flows through an analog or digital channel strip. While this can get quite complicated thanks to the wide array of features on modern consoles and mixers, we can reduce it down to its simple basics. This will illustrate all the basic operations that we'll cover later in the book.

Typical Analog Console Signal Flow

If we look at an analog console first (Figure 1.2), you'll see an example that's pretty typical of most hardware consoles and smaller mixers. The input signal from either a microphone or recorder comes in from the top and first sees a *Trim* or *Gain* control (same thing) which allows you to add or subtract level as needed.

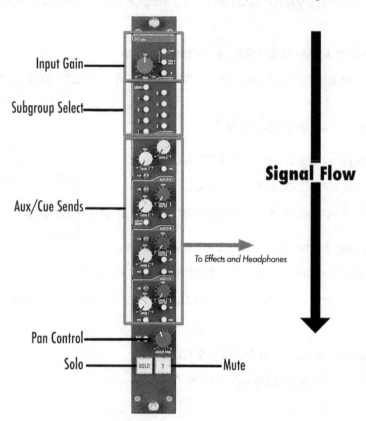

Figure 1.2: An Typical Analog Console Channel Strip

Input Gain

Subgroup Select

Signal Flow

Aux/Cue Sends

To Effects and Headphones

Pan Control

Solo — Mute

Next on this example is the *Subgroup Select* buttons that allow us to assign the signal coming out of the channel to another fader that will control not only the one we're working on, but as many other channels as you want to assign to it as well. For instance, all drum channels could be assigned to one subgroup so you only have to move that one subgroup fader to control the level for them all.

Then we come to a series of controls that are called sends, auxes, aux sends or cue sends (all the same thing). They allow us to send some of the signal coming into the channel to effects like reverb or delay, or to musician's headphone mixes. When sent to headphones (or stage monitors in a live environment) these controls are called "cue sends," but they're exactly the same as an aux send and both can be used interchangeably for either purpose.

Then the signal flows to the controls that you're probably already familiar with; the *Pan* control that adjusts where the signal will be heard across the stereo soundfield; the channel *Solo* control that allows us to listen to only that one channel while muting the rest; and the *Mute* control that kills the signal in the channel so it can't be heard. After that the signal flows to the channel fader (not shown in the example) and then typically to the mix buss (sometimes called the master buss, output buss, 2 mix, stereo buss - all the same) where we can control the overall level of the mix.

What we see in Figure 1.2 is what's known as a Channel Strip and will be repeated dozens of times on a typical recording console. That's what makes large-scale recording consoles that contain 24, 48, 64 or more channels so intimidating at first, but when you remember that each channel strip is the same as all the rest it makes it a lot less daunting.

Typical DAW Mixer Signal Flow

The digital version of an analog console found in a typical DAW works the same but has a lot more flexibility, which can be confusing at first. At the top of the channel comes a series of virtual slots called *Inserts* where we can add digital processing like equalization, compression, or even effects plugins. On the example in Figure 1.3 you'll see a *Trim* plugin that's doing the same job as the *Gain* control seen at the top of Figure 1.2. Depending upon the DAW, this may or may not be automatically added to a channel when you open the DAW session for the first time.

Figure 1.3: A Typical DAW Channel Strip

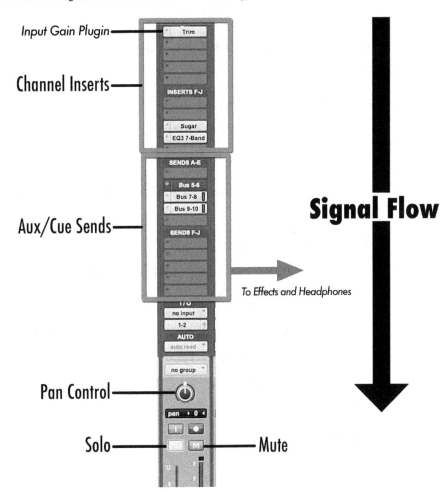

Next comes a number of slots that allows us to insert *Auxes* to send to effects or headphones. Please note the number of slots varies between DAWs, but there are usually at least 4 available.

Finally comes our *Pan, Solo* and *Mute* controls just like on the hardware console, after which the signal flows into the channel fader just like on the hardware console.

Regardless of what kind of console or mixer that you use, the signal flow is always the same. It starts at the top and runs down the channel to the channel fader at the bottom. In between you have a chance to insert signal processors, or split the signal off to send to effects or headphones (or in-ear of floor monitors in sound reinforcement).

TIP: The signal always flows from top to bottom, regardless of if it's a hardware console or in a DAW.

▶ Exercise 1.2: Following The Signal Flow Of Your DAW Channel

Consult your DAW manual or Quick Start to familiarize yourself with its functions first.

A) Locate the mixer page on your DAW

☐ Is it filled with channels or do you have to create them?

☐ Is the processing built-in or do you have to insert it?

B) Locate the Insert section of a channel on your DAW.

☐ Insert a gain or trim plugin.

☐ Insert an EQ plugin (any will do).

☐ Insert a compressor plugin (any will do).

☐ Be familiar with how to bypass the plugins if needed.

C) Locate the Sends section of a channel of your DAW.

☐ Insert a send from buss 1.

☐ Insert a send from buss 2.

☐ Notice the level control on each aux send.

☐ Does the send have a mute button?

Chapter 2
Monitoring

You can have the best monitoring chain that money can buy, but if it isn't set up correctly in your room, your mixes are going to be disappointing. On the other hand, even if you have inexpensive monitor speakers, you can get surprisingly good results if they're placed correctly in the room. *You can't mix it if you can't hear it,* and that's why you have to tighten up your listening environment before you begin to mix.

The Listening Environment

Unless you're purpose-building your studio from the ground up, it's easy to overlook your listening environment. Usually it's just the old "throw some speakers and a DAW up in an open corner of the room and go" routine where we try to get everything making sounds as quickly as possible and leave it at that.

While it's possible that you can get lucky with a balanced sound and wide stereo field by just setting up a couple of nearfield monitor speakers in your room without thinking much about it, usually that's not the case because normal garages, living rooms and bedrooms aren't intended as listening spaces and have little in the way of acoustic treatment. Whether you're treating your room or not (you really should - Read my book *The Studio Builder's Handbook* for inexpensive ways to do it), the following steps are necessary to optimize what you're hearing.

The correct placement of the speakers is one of the most critical adjustments that you can make in improving the sound of your room. Before you do anything else, this place must be determined.

TIP: Sometimes, even just a movement of a few inches backwards or forwards can make a big difference.

Determining The Listening Position

The first thing that to do is to select the best place in the room for your listening position. The place that provides the best acoustic performance will almost always come from setting up lengthwise in the room because it's easier to avoid some of the problem room reflections (see Figure 2.1) that can plague the frequency response of the room. In other words, *the speakers should be firing the long way down the room.*

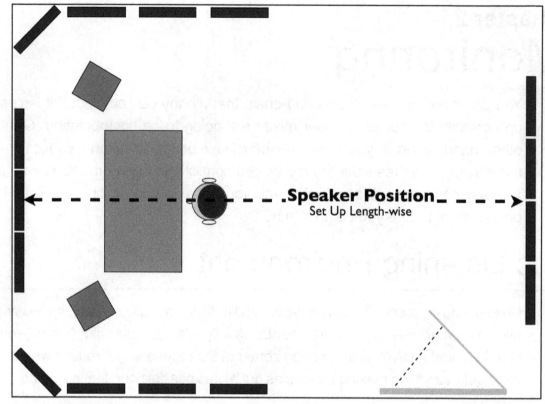

Figure 2.1: Speaker placement in the long part of the room

Acoustic Quick Fixes

Without having to acoustically treat your room (which should always come first if you can), here are a few simple things that you can do to instantly improve the performance of your playback system.

- **Avoid placing the speakers up against a wall.** This usually results in some strong peaks in the low frequency response, which compromises the sound. The further away you can get from the wall, the less it influences the frequency response of your monitors and the smoother that response will be. Figure an absolute minimum of twelve inches, although more is better, as long as you stay out of the 25% point of the room as mentioned above.

TIP: 38% is actually the ideal monitor placement point in the room, but anything other than the 25% or 50% point will work.

- **Avoid the corners of the room.** Even more severe than the wall is a corner, since it will reinforce the low end even more than when placed against a wall. The worst is if only one speaker is in the corner, which causes the response of your system to be lopsided on the low-end towards the speaker located there.

- **Avoid being closer to one wall of the room than the other.** If one speaker is closer to one side wall than the other, once again you'll get a totally different frequency response between the two because of phase and reflection issues. It's best to set up directly in the center of the room if possible. Symmetry is essential to keep a balanced stereo image with a stable frequency response in the room. That means that your listening sweet spot will be in the exact center of the room if the speakers are exactly the same distance from each side wall. While it may seem tempting to set up some other way, acoustically you could be asking for trouble.

- **Avoid different types of wall treatment.** If one side of the room contains a window and the other is drywall, carpet or acoustic foam, once again you'll have an unbalanced stereo image because one side will be brighter sounding than the other. Try to make the walls on each sides of the speakers the same material.

- **Isolate the speakers.** Speakers mounted directly on a desk or console will defeat the purpose of much of the acoustic treatment. Mark the position of the speakers with masking tape, and mark the position in one inch increments up to six inches either way from the wall so you don't have to remeasure in the event that you have to move things. Exact distances are critical, so always use a tape measure because even an inch can make a big difference in the sound.

> **Exercise 2.1:** *Placing Your Speakers In The Room*

A) **Play a song that you think sounds great and you're very familiar with. Place your monitors at the 25% point of your room.**

☐ Does the frequency response change?

☐ Are some bass frequencies missing?

☐ What happened to the stereo image?

☐ Are some bass frequencies reinforced?

B) **Move the speakers a few inches backwards from the 25% point.**

☐ Does the frequency response change?

☐ Are some bass frequencies missing?

☐ Are some bass frequencies reinforced?

☐ Is the response smoother?

C) **Move the speakers a few inches forwards from the 25% point.**

☐ Does the frequency response change?

☐ Are some bass frequencies missing?

☐ What happened to the stereo image?

☐ Are some bass frequencies reinforced?

☐ Is the response smoother?

D) Now move the speakers to the exact center of the room between the walls.

☐ Does the frequency response change?

☐ What happened to the stereo image?

☐ Are some bass frequencies missing?

☐ Are some bass frequencies reinforced?

☐ Is the response smoother?

E) Now move the speakers to the 38% point in the room, if possible.

☐ Does the frequency response change?

☐ Are some bass frequencies missing?

☐ What happened to the stereo image?

☐ Are some bass frequencies reinforced?

☐ Is the response smoother?

F) Now move the speakers so they're against the front wall.

☐ Does the frequency response change?

☐ Are some bass frequencies missing?

☐ What happened to the stereo image?

☐ Are some bass frequencies reinforced?

☐ Is the response smoother?

G) Place one speaker in a corner of the room.

☐ Does the frequency response change?

☐ Are some bass frequencies missing?

☐ Are some bass frequencies reinforced?

☐ Is the response smoother?

H) Now move the speakers so they're 12 to 18 inches away from the front wall.

☐ Does the frequency response change?

☐ Are some bass frequencies missing?

☐ What happened to the stereo image?

☐ Are some bass frequencies reinforced?

☐ Is the response smoother?

Basic Monitor Setup

Now that your listening position is placed correctly in the room, it's time to set up your monitors. While most home studios seem to have a random amount of space between their monitors, there are a number of general guidelines you can use to optimize your setup. Since most rooms are unique in some way in terms of dimensions or absorbent qualities, you may have to vary from the following outline a little, but these are good places to start from.

- **Check The Distance Between The Monitors -** If the monitors are too close together, the stereo field will lack definition. If the monitors are too far apart, the focal point or "sweet spot" will be too far behind your head and you'll hear the left or the right side individually, but not both together as one. The rule of thumb is that the speakers should be as far apart as their distance from the listening position. That is, if your listening position is four feet away from the monitors, then start by moving them four feet apart so that you make an equilateral triangle between you and the two monitors (Figure 2.2).

TIP: It's been found that 67 ½ inches from tweeter to tweeter seems to be an optimum distance between speakers, and focuses the speakers three to six inches behind your head (which is exactly what you want).

Figure 2.2: Monitor speakers and listener in an equilateral triangle

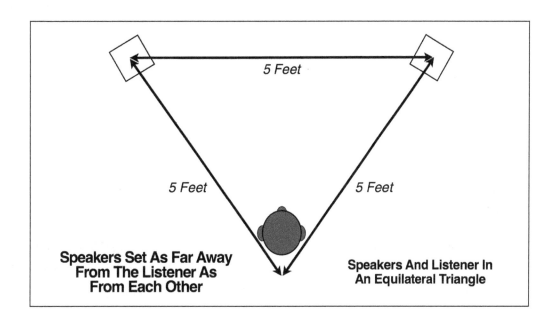

- **Check The Angle Of The Monitors -** Not angling the speakers properly will cause smearing of the stereo field, which is a major cause of a lack of instrument definition when you're listening to your mix. The correct angle is somewhat determined by taste, as some mixers prefer the monitors angled directly at their mixing position while others prefer the focal point (the point where the sound from the tweeters converges) anywhere from three to twenty-four inches behind them to widen the stereo field (see Figure 2.3).

TIP: *It's been found over time that an angle of 30 degrees that's focused about 18 inches behind the mixer's head works the best in most cases.*

Figure 2.3: The monitor focal point

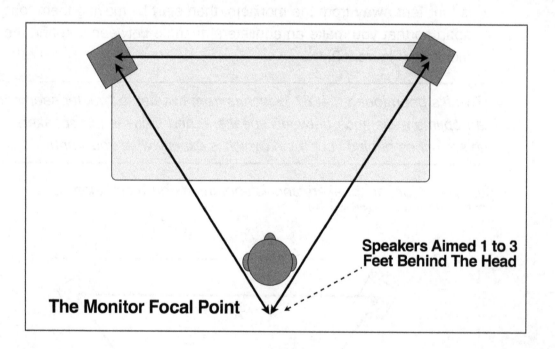

Speakers Aimed 1 to 3 Feet Behind The Head

The Monitor Focal Point

A great trick for finding the correct angle is to mount a mirror over each tweeter and adjust the speakers so that your face is clearly seen in both mirrors at the same time when you are in your mixing position.

- **Check How The Monitors Are Mounted** - If at all possible, it's best to mount your monitor speakers on stands just directly behind the meter bridge of the console or edge of your desk. This gives you a much smoother frequency response.

Monitors that are placed directly on top of a computer desk or console meter bridge without using any isolation are subject to low frequency cancellations because the sound travels through the desk or console, through the floor and reaches your ears before the direct sound from the monitors through the air gets there. This causes some frequency cancellation and a general smearing effect of the audio. If you must set your speakers on the desk or console, place them on a 1/2 or 3/4 inch piece of open cell neoprene, a thick mouse pad or two, or something like the Prime Acoustic Recoil Stabilizers (see Figure 2.4). You'll be surprised how much better they sound as a result.

Figure 2.4: Yamaha NS-10 on a Primacoustic Recoil Stabilizer
Courtesy of Primeacoustic

- **Check How The Monitor Parameters Are Set** - Almost everyone uses powered monitors these days, but don't forget that many have a few parameter controls either on the front or rear of the speaker. Be sure that these are set correctly for the application (make sure you read the manual) and *are set the same on each monitor.*

- **Check The Position Of The Tweeters** - Many monitors are meant to be used in an upright position, yet users frequently will lay them down on their sides. That makes them easier to see over, but the frequency response suffers as a result. That being said, if the speakers are designed to lay on their sides, most mixers prefer that the tweeters be on the outside towards the walls because the stereo field is widened (see Figure 2.5). Sometimes tweeters to the inside works but that usually results in the stereo image smearing. Try it both ways and see which one works best for your application.

If your speakers are placed upright, be sure that the tweeters are head-height since the high frequency response at the mixers position will suffer if they're too high and firing over your head. Sometimes it's necessary to even flip them over and place them on their tops in order to get the proper tweeter height.

Figure 2.5: Speakers with tweeters to the outside

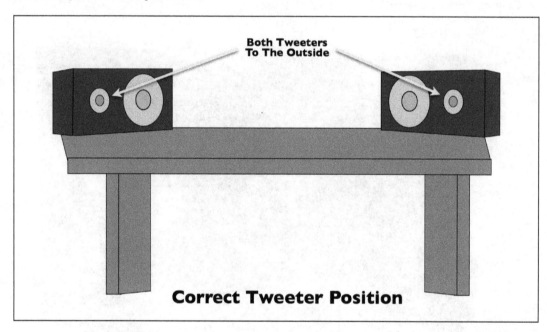

▶ *Exercise 2.2: Spacing Your Speakers*

A) Place the monitors exactly 67 1/2 inches apart from tweeter to tweeter. Play a song that you think sounds great and you're very familiar with.

☐ What does the stereo image sound like?

☐ What happens to the frequency balance of the speakers?

B) Now move the speakers closer together.

☐ What happens to the stereo image?

☐ What happens to the frequency balance of the speakers?

C) Now move the speakers further apart beyond the 67 1/2 inches.

☐ What happens to the stereo image?

☐ What happens to the frequency balance of the speakers?

> ### Exercise 2.3: *Angling Your Monitor Speakers*

A) Angle the monitors so that they are pointing straight at the back of the room.

☐ What is the stereo image like?

B) Angle the monitors so that they are pointing directly at the center of your head.

☐ What is the stereo image like?

C) Now angle the monitors so that they're aiming at a point about six inches behind your head.

☐ What is the stereo image like now?

> ### Exercise 2.4: *Isolating Your Monitor Speakers*

A) If your monitors are sitting on your desk or console, place a mouse pad or two underneath each one.

☐ Did the sound change?

☐ Did the low end get tighter and more focused?

☐ What's the stereo field like?

B) Place the speakers back on the console or desk, but without the mouse pads.

☐ Does it sound any different?

☐ What's the stereo field like?

C) If you have stands, now place them back on the stands.

☐ Does it sound any different?

☐ What's the stereo field like?

> ### Exercise 2.5: *Setting The Speaker Parameters*

A) If your monitor speakers have parameter controls on the back, randomly change the controls on the right speaker.

☐ What happens to the frequency response?

☐ What happens with the stereo image?

B) Now set the parameter controls so that they're exactly the same on each speaker (including the volume control if there is one).

☐ What happens to the stereo image?

☐ What happens to the low-end frequency response?

▶ Exercise 2.6: Positioning The Speaker Tweeters

A) Lay each speaker down horizontally on its side so that the tweeters are positioned toward the wall.

☐ What happens to the stereo image?

☐ What happens to the frequency response?

B) Now flip each speaker so that the tweeters are positioned on the inside towards each other.

☐ What happens to the stereo image?

☐ What happens to the frequency response?

How To Listen

It's time to develop your critical listening skills. We've been listening to the world around us all of our lives without thinking about it, but when it comes to mixing, we must take our listening ability to another level all together. The following explanations and exercises are designed to make you think about what you're listening to while mixing, as well as provide you with set of tools to make sure that your mix translates well to all listening environments.

Basic Listening Technique

All of your life you've been listening to things as a whole. When you were outside in the park, you heard the dogs barking, birds chirping, a police car siren in the distance, children laughing while they're playing; but you've mostly heard it all as one sound and only occasionally zeroed in on a portion of it without realizing it. When you went to a party, you heard the music in the background, guests laughing and talking, ice cubes tinkling in the glass as drinks are being made, and your ears might've picked up a conversation in the distance with a tidbit of juicy gossip. That was the beginning of critical listening.

When you listened to a song on the radio, you listened to the song as a whole, and never really listened to the individual parts of the mix, unless you heard a particular instrument like the one you play.

Now it's time to begin to break those sound sculptures down into individual parts. This means identifying individual musical instruments, mix elements, frequency response, mix balance, and ambience, to name a few. Beware, after you begin to do this, you'll never hear music in the same way again, for better or worse.

► Exercise 2.7: *What To Listen For*

Play one of your favorite songs, but make sure it's available at the highest quality resolution possible, which means CD, vinyl, high-resolution file, or one of your mixes directly from your DAW. You're going to learn to listen to this mix in a different way.

A) Listen to the mix.

☐ How many individual instruments can you identify?

B) Listen to the mix again.

☐ What part of the frequency spectrum does each instrument or vocal take up?

C) Listen to the mix again.

☐ How many different types of ambience (natural or artificial reverb) can you identify?

☐ Does each mix element have its own ambience?

☐ Is there one general ambience that that seems like it's on everything?

☐ Does anything have a noticeable delay?

D) Listen to the mix again.

☐ What is the most interesting thing in the mix?

☐ Is it a vocal or an instrument or a sound effect?

☐ Is it a riff or lead line?

☐ Is there an effect that catches your ear?

E) Listen to the mix again.

☐ What is the frequency balance of the mix as a whole like?

☐ Are there a lot of high frequencies?

☐ Are the mids prominent?

☐ Does the song have a deep bottom end?

F) Listen to the mix again.

☐ Does the song have a lot of dynamics?

☐ Does it seemed to be compressed?

☐ Can you hear the compression on individual instruments or vocals?

☐ Is the compression pleasing to the ear or difficult to listen to?

How Loud (Or Soft) Should I Listen?

One of the greatest misconceptions about music mixers (especially the great ones) is that they mix at high volume levels. Some do, and at excruciatingly loud levels as well, but most mixers find that they get better balances that translate well to the real listening world by monitoring at conversation level or even lower.

High sound pressure levels for long periods of time are generally not recommended for the following reasons:

1) First the obvious one, exposure to high volume levels over long periods of time my cause long-term physical damage.

2) High volume levels for long periods of time will not only cause the onset of ear fatigue, but physical fatigue as well. This means that you might effectively only be able to work six hours instead of the normal eight (or ten or twelve) that's possible if listening at lower levels.

3) The ear has different frequency response curves at high volume levels that overcompensate on both the high and low frequencies. This means that your high volume mix will generally sound pretty limp when it's played at softer levels.

4) Balances tend to blur at higher levels. What sounds great at higher levels won't necessarily sound that way when played softer. However, balances that are made at softer levels always work when played louder.

Now this isn't to say that all mixing should be done at the same level and everything should be played quietly. In fact, music mixers (as opposed to film mixing, which always uses one constant level) tend to work at a variety of levels; up loud for a minute to check the low end, and moderate while checking the EQ and effects. But the final balances usually will be done quietly.

TIP: Sometimes, the only way that you can check the low frequencies on a mix is to turn it up to a moderately loud level for a brief period, so don't be afraid to do that if needed. Just remember that keeping it up loud for long periods of time probably won't help your mix translate to other playback systems too well.

▶ Exercise 2.8: Setting The Listening Levels

A) Listen to a mix at low level, then turn it up to a medium level (comfortable), then a very loud level.

☐ Can you hear the frequency response changing as the volume changes?

☐ Is there more or less high end?

☐ Is there more or less low end?

B) Listen to the mix at a very low, barely perceptible level.

☐ Is there a mix element that jumps out?

☐ Is there a mix element that disappears?

C) Increase the level of the kick drum by 2dB.

☐ Can you hear the balance of the song change?

☐ Does the kick drum seem louder?

D) Return the kick drum to its original level. Now increase the level of the bass element by 2dB.

☐ Can you hear the balance of the song change?

☐ Does the bass element seem louder?

☐ If you can't hear the difference at 2dB, at what level can you hear the difference at?

E) Now turn the level of the mix up to a level that's louder that comfortable. Increase the level of the kick drum by 2dB.

☐ Can you hear the balance of the song change?

☐ Does the kick drum seem louder?

F) Return the kick drum to its original level. Now increase the level of the bass element by 2dB.

☐ Can you hear the balance of the song change?

☐ Does the bass element seem louder?

☐ If you can't hear the difference at 2dB, at what level can you hear the difference at?

G) Listen to your mix at a very low, barely perceptible level. Raise the level of the lead vocal by 2dB.

☐ Can you hear the balance of the song change?

☐ Does the vocal seem louder?

H) Return the lead vocal to its original level. Now increase the level of the rhythm guitar by 2dB.

☐ Can you hear the balance of the song change?

☐ Does the guitar seem louder?

☐ If you can't hear the difference at 2dB, at what level can you hear a difference at?

I) Now turn the level of the mix up to a level that's louder that comfortable. Increase the level of the lead vocal by 2dB.

☐ Can you hear the balance of the song change?

☐ Does the vocal seem louder?

J) Return the lead vocal to its original level. Now increase the level of the rhythm guitar by 2dB.

☐ Can you hear the balance of the song change?

☐ Does the guitar seem louder?

☐ If you can't hear the difference at 2dB, at what level can you hear the difference at?

Listening On Several Speaker Systems

If you don't have an alternate monitor system yet, then what are you waiting for? Most veteran mixers use at least a couple of systems to get a feel for how everything sounds - the main system where the mixer does all of the major listening work, and an alternate system for a different perspective.

The alternate speaker is used simply as a balance check to make sure that one of the instruments isn't either too loud or too soft in the mix. Also, one of the arts of mix balance is getting the kick drum and bass guitar to speak well on a small system, which is why an alternative monitor system is so important.

The second set of monitors doesn't have to be great. In fact, the crappier they are, the better. Even a set of ten dollar computer speakers can work. The idea is to have a second set that will give you an idea of what things sound like in that world, since unfortunately, there are a lot more people listening on crappy monitors than good ones these days.

▶ **Exercise 2.9:** *Listening On Multiple Monitors*

A) Listen to a mix on your normal monitors, then switch to your alternate monitors (make sure they're at exactly the same level).

☐ Is the frequency response different?

☐ Are the mids emphasized, or are they attenuated a bit?

☐ Is the balance different?

☐ Is the low end different?

☐ Can you hear the bass element better, or does it disappear?

☐ Can you hear the kick drum better, or does it disappear?

Listening In Mono

Sooner or later your mix will be played back in mono somewhere along the line, so it's best to check what will happen before you're surprised later. Listening in mono is a time-tested operation that gives the mixer the ability to check phase coherency and balances. Let's look at each one individually.

Phase Coherency

When a stereo mix is combined into mono, any elements that are out of phase will drop in level or even completely cancel out. This could be because the left and right outputs to the speakers are wired out-of-phase (pin 2 and pin 3 of the XLR connector are reversed) which is the worst-case scenario, a cable was similarly out-of-phase when recording, or perhaps because a phase switch was inadvertently selected during recording or mixing.

Regardless of how it happened, an out of phase effect causes the lead vocal or instrument solo to cancel out and disappear from the mix, which certainly isn't something that you want to have happen. As a result, it's prudent to listen in mono once in a while just to make sure that a mono disaster isn't lurking in the wings.

Balances

Many engineers listen to their mix in mono strictly to balance elements together since they feel that they hear the balance better this way. Listening in mono is also a great way to tell when an element is masking another. As legendary engineer Andy Johns (Led Zeppelin, the Rolling Stones, Van Halen, Eric Clapton) once told me, "That used to be the big test (mixing in mono). It was harder to do and you had to be a bloody expert to make it work. In the old days we did mono mixes first then did a quick one for stereo. We'd spend 8 hours on the mono mix and half an hour on the stereo."

 Exercise 2.10*: Listening In Mono*

A) If there is a mono switch on your console or DAW, select it to listen to the mix in mono.
☐ Have any instruments, vocals or reverb disappeared from the mix?

B) Listen to the mix in mono again.
☐ Are there any instruments that are fighting or covering another instrument or vocal up?

C) Listen to the mix in mono. Raise the level and walk out of the room and listen from afar.
☐ Does the mix still hold together or does one instrument, vocal or frequency stand out?

◆

Chapter 3
Session Setup

There's a good argument to be made that the setup for a mix session is almost as critical as the mix itself. Proper setup allows you to get into the right headspace to hear what you need to hear over long periods of time. It helps your efficiency during the mix when you know the session is properly labeled and all the assignments, effects and routing are preset beforehand. Once you get into the flow of things, you don't ever want to stop for something technical that could've been taken care of a lot earlier.

We can break session setup down into two elements; prepping your mix and prepping yourself. Let's look at each.

Prepping Your Mix

This is where you get everything in your session prepped for the mix by making things easy to find. You may have to tweak things as you go along, but it will take you far less time as a result of this process, which will keep you in the creative flow of your mix.

Make A Session File Copy

Before you do anything else, make a copy of the session file that's designated as the "mix" and name it something descriptive like "songtitle mix" so it's easy to locate (see Figure 3.1). This also keeps your previous session safe if you ever have to go back to it.

Figure 3.1: A descriptive file names

Name		Date Modified	∨	Size	Kind
	Never Like mix Automate.ptx	Today at 7:50 PM	◉	325 KB	Pro To...Session
	Never Like Example.ptx	Today at 7:49 PM		324 KB	Pro To...Session
▶	Session File Backups	Today at 4:58 PM		--	Folder
	Never Like Vox Start.ptx	Jun 28, 2020 at 3:22 PM		310 KB	Pro To...Session
	Never Like B&D Start.ptx	Jun 28, 2020 at 1:55 PM		306 KB	Pro To...Session
	Never Like B&D.ptx	Jun 28, 2020 at 12:17 PM		306 KB	Pro To...Session
	Never Like DC.ptx	Jun 26, 2020 at 7:54 PM		186 KB	Pro To...Session
	Never Like New Plugs.ptx	Jun 17, 2020 at 6:27 PM		230 KB	Pro To...Session
	Never Like Start.ptx	Dec 7, 2019 at 5:35 PM		126 KB	Pro To...Session
	WaveCache.wfm	Dec 7, 2019 at 5:28 PM		4.2 MB	Overview Cache

I personally always put a date in the file name, but that's not necessary since most of the time it's built into the meta data and can be easily determined by looking at the file info. It's not uncommon to have multiple versions of the same session during the same day, so I like to differentiate one from another with letters of the alphabet at the end of the tittle, like "rosegarden mix 9-9-20a", "rosegarden mix 9-9-20b" and so on.

If you are able, color code the file as well so it's easier to identify. I like to start with a series of colors that show the stage of completion. Like a traffic light, I'll start with red for "stop" and end with green for "go" for finished, but obviously use whatever colors work for you.

TIP: While you're at it, make a copy of the session file on another hard drive, flash drive, online backup, or any place that you can easily grab it if for some reason you find the file you're working on is suddenly corrupted.

Arrange Your Tracks

Deleting or hiding tracks that won't be used and ordering the tracks that will be used may be the single most useful thing you can do while prepping your mix. Here's what to do.

Delete Empty Tracks

Any empty tracks take up space in your edit and mix windows without adding anything useful, so it's best to delete them. During tracking or overdubs, it frequently makes sense to have empty tracks readily available, but if you've gotten to the mix without using them, you know they probably won't be needed. Delete them.

Deactivate And Hide Unused Tracks

Any tracks that you know won't be used just soak up your computer's system resources. Even if you have a power machine, these resources may become a precious quantity if you end up using a lot of plug-ins during the mix. Deactivate them, then hide them from the timeline and mix panels so they don't distract you.

Reorder Your Tracks

This isn't absolutely necessary, but it does make tracks easier to find during the mix. The idea is to group any similar instruments or vocals together, so all the guitars are next to each other, the drums and percussion next to one another, and all the vocals are together.

Color-Code The Tracks

Once again this isn't absolutely necessary, but it sure does make things easier to find if your DAW app has this ability. For instance, all the drums might be red, guitars blue, the vocals yellow, and so on.

Correctly Label The Tracks

Many workstation apps automatically assign a name to any new track that has been recorded, but unfortunately they usually don't relate to the instrument. It's really easy to mistake one track for another and turn a fader or parameter knob up and up and wonder why nothing is happening, only to find that you're tweaking the wrong track. That's why it's important to clearly label each track. You'll be happy with yourself later if you relabel a track with a name like "gt166," or "Dave," to something like "lead guitar" or "ld gtr."

Insert Section Markers

Markers are truly one of the big time savers in any DAW, and if you haven't done so already, now is the time to do it. Most veteran mixers insert a marker a bar or two before each new section, and also make sure that other points like drum fills, accents or even the half-way point in a section are marked as well (see Figure 3.2).

Figure 3.2: Section markers

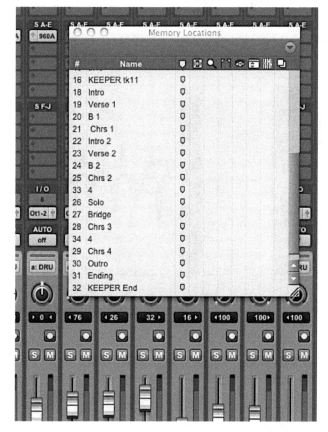

Cleaning Your Tracks

Now is the time to eliminate noises from each individual track. Although they might not sound too bad when all the other tracks are mixed in, after everything is mixed and mastered you'd be surprised by how something that was once buried can now come to the forefront and bother you. Also, by eliminating any extraneous noises, all the tracks magically sound more distinct and uncluttered. Remember, noise is cumulative.

- **Trim the heads and tails.** Trim all the extra record time at the beginning and end of each track, regardless of whether it was recorded during basics or overdubs. Add a fade-in and fade-out to eliminate any edit noise (see Figure 3.3).

- **Crossfade your edits.** One of the biggest problems for A-list mixers when they get a session that's full of edits is that the edits click and pop because they don't contain any crossfades. Even if you can't hear a click or pop, it's a good practice to have a short crossfade on every edit to eliminate the possibility of an unwanted noise (see Figure 3.3). Again, these noises are cumulative!

- **Delete extra notes from MIDI tracks.** Delete any extra "split" notes that were mistakenly played. You might not hear them when all the instruments are playing, but just like the noise at the beginning of tracks, they have a tendency to come to the forefront after things get compressed.

Figure 3.3 Eliminating track noise

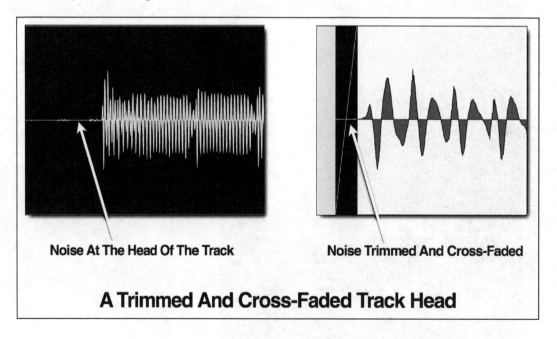

Noise At The Head Of The Track   **Noise Trimmed And Cross-Faded**

A Trimmed And Cross-Faded Track Head

Fix The Timing

Before mixing it's important to fix any timing issues that you might find in your tracks. Prior to mixing these mistakes might not seem obvious, but after all the tracks are tweaked during the mixing process some timing errors that you didn't notice before will now stick out like a sore thumb. It's best to fix them now before the mix even begins.

In the graphic below (Figure 3.4) you can see an example of a track that was played ahead of the beat, and how it was fixed with an edit.

Figure 3.4 Fixing timing issues

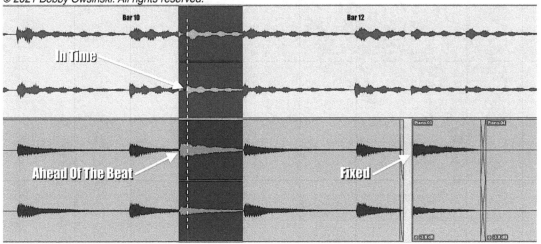

Set Up Groups Or Subgroups

Groups or subgroups are extremely useful during mixing because they allow you to group similar elements of the mix so you can make adjustments by instrument sections, rather than individually (see Figure 3.5). Your mix will go a lot faster if your set up the groups or subgroups and assign the particular channels to them ahead of time.

The difference between them is that a group links all the assigned faders together. When you adjust one, they all move yet stay at the same relative balance. A subgroup assigns all of the audio from the assigned faders to a submaster fader. You can change the level with the subgroup fader but the individual assigned faders don't move. You can process all the channels by just inserting the process on just the subgroup channel, saving time and computer resources.

Typical groups might be drums, guitars (if there's more than one or they're in stereo), lead Vocals (if there's a double), Background vocals, horns, strings And synths.

Set Up Your Effects

Most mixers have a standard set of effects that they set up before they mix. We'll cover this more in Chapters 8 and 9, but one setup that works well even for tracking and overdubs is:

- For drums - A reverb using a dark room set to about 1.2 seconds of decay with a pre-delay of 20 milliseconds

- For all other instruments - A plate with about 1.8 seconds of decay and a pre-delay of 20 milliseconds

- For Vocals - A delay of about 220 milliseconds

It's amazing how well these settings work without any tweaking. Another common setup is two reverbs and two delays, and they're set like:

- Short Reverb - A room program with the decay set from 0.5 to 1.5 seconds of decay with a short pre-delay timed to the track

- Long Reverb - A plate or hall program with a decay set from 1.5 to 4 seconds of decay and a pre-delay of as little as 0ms or as much as 150ms timed to the track (depends on your taste and what's right for the song - we'll cover this in a later chapter)

- Short Delay - A delay of about 50 to 200 milliseconds

- Long Delay - A delay from about 200 to 400 or milliseconds

Your own particular starting point might use a lot more effects, or you may prefer to add effects as the needed during the mix. Regardless, it's a good idea to have at least some effects set up before you start the mix so you won't break your concentration to set them up later. We'll look at some other effects setups later in Chapters 8 and 9.

Assign The Channels

Usually you know ahead of time that there are certain tracks that will use a certain effect (like the drums or snare to a short reverb). It's best to assign those channels to the appropriate sends and pan them accordingly before your mix begins.

Insert Compressors And Limiters

Chances are that at least a few channels, like the kick, snare, claps, bass and vocal, will need a compressor during the mix. That's why it's a good idea to insert the compressor ahead of time during mix prep. Just remember to leave it bypassed until you decide you need it (see Figure 3.3).

▶ *Exercise 3.1: Prepping Your Mix*

A) Make a copy of the session file.

☐ Rename it and assign the file a color so it's easy to identify.

☐ Can you easily find it in a group of files?

B) Make a second copy of the session file on another hard drive, flash drive, or online backup as a safety.

_____

▶ *Exercise 3.2:* Arranging Your Tracks.

A) Delete empty tracks.

B) Deactivate and hide any tracks that won't be used in the mix.

C) Reorder the tracks by instrument, loop, sample, beat, or vocal kind.

D) Color code the tracks in logical groups.

E) Rename the tracks so they provide a description of the track and so they're easy to find.

▶ *Exercise 3.3:* Creating Markers

A) Go through the song and add markers right before the beginning of every section.

☐ Can you easily find them?

☐ Are they named appropriately?

B) Go through the song and add markers right before noteworthy spots like sound effects, fills, and B sections.

☐ Can you easily find them?

☐ Are they named appropriately?

▶ *Exercise 3.4:* Eliminating Extraneous Track Noise

Use the "It's About Time" example song, since there are many tracks that require cleanup.

A) Solo the lead Vocal track and listen before the vocal starts (before the waveform gets large).

☐ Can you hear noise, coughs, breathing, etc.? If there's no noise, proceed to another track.

B) If there's noise, grab the edge of the clip and move it towards where the vocal begins (refer to Figure 3.3).

☐ Can you still hear the noise when played back?

☐ Can you still hear the attack of the first phrase of the vocal? If not, extend the clip up a bit so the attack is heard cleanly.

C) Alternatively, make at edit just at the start of the waveform of the vocal. Highlight the separated clip on the left and delete (refer to Figure 3.3).

☐ Can you still hear the noise when played back?

☐ Can you still hear the attack of the first phrase of the vocal? If not, extend the clip up a bit so the attack is heard cleanly.

D) Place a short fade where the clip now starts. This will eliminate any pops or clicks as a result of the edit (refer to Figure 3.3)..

☐ Can you still hear the noise when played back?

☐ Can you still hear the attack of the first phrase of the vocal? If not, extend the clip up a bit so the attack is heard cleanly.

E) Do the same as above where the vocal stops. Be careful not to delete any other vocal part further down the track.

☐ Can you still hear the noise when played back?

☐ Can you still hear the release of the last phrase of the vocal? If not, extend the clip up a bit so the release is heard cleanly.

F) Proceed to all other tracks using the same method as above.

▶ Exercise 3.5: *Fixing Track Timing*

in the example song "It's About Time" there are 4 timing issues on the Piano track - on bars 10, 12, 34 and 36. In all cases the piano is played before the beat instead of right on it like the other keyboards. Go to bar 10 and solo the Piano and Rhodes tracks. Refer to Figure 3.4 to be sure that you're in the right place.

A) Cut right before the waveform that's ahead of the beat, then right behind it but right before the next note. Nudge the clip backwards until it's exactly lined up with the Rhodes track.

☐ Do both tracks sound like they're playing together now?

☐ What happens to the note right behind the edited clip? Try shortening the clip so it doesn't overlap with the next note.

☐ Can you hear any pops or clicks from the edits?

B) Place a short fade at all edit points.

☐ Can you hear any pops or clicks from the edits now?

C) Repeat on bars 12, 34 and 36.

▶ Exercise 3.6: *Assigning Subgroups*

A) Set up and assign subgroups to anything with multiple tracks like drums, guitars, and vocals.

☐ Are the subgroups named appropriately?

☐ Do they work as expected? (Try muting the group to see if all assigned channels are silenced as well.)

▶ Exercise 3.7: Creating Effects Channels

A) Add four effects channels.

B) Insert a reverb on the first two

C) Insert delays on effects channels 3 and 4

D) Assign them to the stereo buss

E) Assign the sends to the effects to the appropriate channels

---

▶ Exercise 3.8: Inserting Channel Processors

A) Insert a compressor on the kick drum channel or subgroup

B) Insert a compressor on the snare drum channel or subgroup

C) Insert a compressor on the bass element channel or subgroup

D) Insert a compressor on the lead vocal channel or subgroup

E) Insert a compressor on the background vocal subgroup

F) Make sure that all processors are bypassed

Prepping Yourself

Now that the technical portion of the mix is set up, it's time to get yourself prepared to mix. Each mix requires focus and concentration, and this is where we get ourselves into the proper headspace.

Play Something You Know

The most important thing during your own personal prep time is to play at least one song or mix that you know well so you have a reference point as to what the room and monitors sound like. Listening to a mix or two will also calibrate your ears to the listening environment, which will help to keep you from over or under-EQing as you go along.

Take Notes

During the course of a long mix you'll probably have to take some notes, so have a pen and a pad of paper, or even some Post-it notes, ready to write on. If you're using a hardware controller, you'll need a roll of console tape.

TIP: Permacel P-724 is the tape type used to mark the names of the channel that can be reapplied without leaving any sticky residue behind.

Make Yourself Comfortable

Most mixes take a while, so you need to be comfortable. Make sure your clothes and shoes are comfy, the room temperature is just right, and the lighting is adjusted so you can easily see any monitor screens that you may be using without any glare. It's also a good idea to have some beverages and a snack ready for later when you need a break.

Setting up for a mix is a lot more work that you might have thought, but it's time put to good use. Once these things are out of the way, your files, tracks, mind and ears are all set for the mix ahead.

▶ Exercise 3.9: Prepping Yourself

A) Play a song or mix that you know well at a medium comfortable volume.

☐ Are you being distracted by any outside noises?

B) Play the same song or mix at a very *loud* volume.

☐ Is the frequency response of the monitors any different?

C) Play the same song or mix at a very *low* volume.

☐ Is the frequency response of the monitors any different?

_____

▶ Exercise 3.10: Prepping Your Work Surface

A) Have plenty of pens, a Sharpie, and blank paper available.

B) Apply console tape and name the tracks if you're using a console or controller.

_____

▶ Exercise 3.11: Prepping Your Environment

A) Make sure that your environment is comfortable.

☐ Are your shoes and clothing comfortable?

☐ Is the room temperature comfortable?

☐ Is the lighting at the correct brightness level?

☐ Is the listening environment quiet?

☐ Do you have refreshments available?

◆

Chapter 4
Mixing Mechanics

Although most engineers ultimately rely on their intuition when doing a mix, there are certain mixing procedures that they all consciously or unconsciously follow.

The Overall Mixing Approach

When asked what you're trying to accomplish when mixing, it's easy to say, "I'm just trying to balance everything together." Sure, that's one aspect of it, but that's not all. If you were to analyze it, mixing comes down to three things:

1. **Developing the groove**

2. **Emphasizing the most important elements**

3. **Putting the performers in an environment**

Let's look at each one.

Developing The Groove

The groove is the pulse of the song. It's that undeniable feeling that makes you want to get off your seat and shake your booty. You don't have to know what it is as much as recognize it when it's there, or when it's not. Despite what you might think, it's not only dance music that has a groove. Every kind of music, whether it's R&B, jazz, rock, country, or some alien space music, has a groove, but the better the music is performed, the "deeper" the groove is.

TIP: Contrary to popular belief, a groove doesn't have to have perfect time because a groove is created by tension against even time. As a result, the playing doesn't have to be perfect, it just has to be even in its execution. In fact, music loses its groove if it's too perfect, which is why a song can sound lifeless after it's been quantized in a workstation. It's lost its groove.

Another misconception is that the groove always comes from the drums and/or bass, but it could be other instruments as well. For instance, the Police's "Every Breath You Take" has the rhythm guitar establish the groove, Ed Sheeran's "Shape Of You" uses a synthesizer line, while most of the Motown hits of the 60s relied on James Jamerson's bass.

Regardless of what instrument is providing the groove of the song, if you want a great mix, you've got to find it and develop it first before you do anything else.

▶ Exercise 4.1: Identifying The Groove

A) In order to hear a groove at its best, let's go to the masters. Play any song by James Brown, Prince, Sly and the Family Stone or George Clinton.

☐ Can you feel the pulse of the song, the groove?

☐ Can you identify the mix elements that are providing the groove?

▶ Exercise 4.2: Identifying The Groove - Part 2

A) Pick one of your favorite songs and have a listen.

☐ Can you feel the pulse of the song?

☐ What mix elements are providing the groove?

B) Play a song at random.

☐ Can you feel the pulse of the song?

☐ What mix elements are providing the groove?

C) Play a song from a genre that you seldom listen to.

☐ Can you feel the pulse of the song?

☐ What mix elements are providing the groove?

▶ Exercise 4.3: Identifying The Groove - Part 3

A) Now listen to all of those songs again.

☐ What makes the groove stand out?

☐ Is it the balance of the mix elements?

☐ Is it because the mix elements providing the groove are louder?

☐ Is it the tone of the mix elements?

☐ Are they punchier sounding than the others?

Emphasizing The Most Important Elements

Every song has some element that acts like a hook to capture a listener's attention. Many times it's the vocal but it can be other elements as well. Usually it's an element that's so important that without it, the song just wouldn't be the same. It could be the piano line in Coldplay's "Clocks," the synthesizer line in Maroon 5's "Move Like Jagger," or the intro guitar line in the Rolling Stone's seminal "Satisfaction," where if they weren't there, it would almost be a different song.

Finding the most important mix element is vital to getting a great mix. It's what provides the excitement and the reason to listen. In a dance song it might be the kick drum, in an R&B song it could be the groove, in a pop song it might be an interesting hook that an instrument plays in the intro and interludes, and yes, it could be the vocal in just about any genre. The thing is, you have to listen to each of the elements to discover exactly what the element is that drives the song. Once that element is found, make sure it's emphasized. You'll see how in the chapters after this.

> **Exercise 4.4**: *Identifying The Most Important Song Element*

A) Pick one of your favorite songs and have a listen.

☐ What mix elements are the most important?

☐ Would the song be different if it wasn't there?

B) Play a song at random.

☐ What mix elements are the most important?

☐ Would the song be different if it wasn't there?

C) Play a song from a genre that you seldom listen to.

☐ What mix elements are the most important?

☐ Would the song be different if it wasn't there?

Putting The Performers In An Environment

Nothing sounds more boring and more like a demo than a flat recording where the musicians sound like they're playing in a closet right next to you. Not only is it unnatural sounding, but it's usually not that exciting to the ear. If you listen to any concert or live music in a club, it's the ambience of the space that makes it come alive.

That's why it's important to learn to how to put each element into its own environment, either by placement when recording or artificially when mixing. While it's easy to think that we're just talking about reverb and delay here, that's not the case. It's the concept of "tall, deep, and wide" (see Figure 4.1).

Figure 4.1: Tall, Deep, And Wide

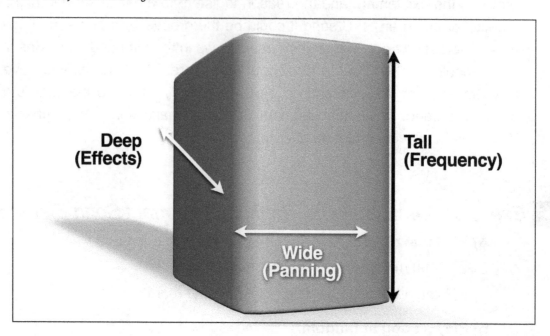

In order to do a great mix you must think in three dimensions: "tall, deep and wide." This means that all the frequencies of the audio spectrum are represented, the mix has some ambient depth, and it has some stereo width from left to right.

• The "Tall" or frequency dimension comes from knowing what frequencies are missing or are too predominant, which means that all of the sparkly, tinkly highs and fat, powerful lows are there, and that all instruments can be distinctly heard.

• The "Deep" or effects dimension is achieved by introducing new ambience elements into the mix. This is usually done with reverbs and delays (and other effects like flanging and chorusing), but room mics, overheads and even leakage play an equally big part as well.

• The "Wide" or panning dimension comes from placing a musical element in a sound field in such a way that it becomes a more interesting soundscape, and as a result, each element is heard more clearly.

> ## Exercise 4.5: *Identifying The Tall, Deep And Wide Dimensions*

A) Pick one of your favorite songs and have a listen.

☐ What mix elements contain the highest frequencies?

☐ What mix elements contain the lowest frequencies?

☐ What mix elements are mostly mid-range?

☐ What mix elements have the most ambience?

☐ Does it seem to have a wide stereo field or is it closer to mono?

B) Play a song at random.

☐ What mix elements contain the highest frequencies?

☐ What mix elements contain the lowest frequencies?

☐ What mix elements are mostly mid-range?

☐ What mix elements have the most ambience?

☐ Does it seem to have a wide stereo field or is it closer to mono?

C) Play a song from a genre that you seldom listen to.

☐ What mix elements contain the highest frequencies?

☐ What mix elements contain the lowest frequencies?

☐ What mix elements are mostly mid-range?

☐ What mix elements have the most ambience?

☐ Does it seem to have a wide stereo field or is it closer to mono?

The 6 Elements Of A Mix

Every genre of music that has a strong backbeat has six main elements to a great mix. They are:

Balance - the volume level relationship between musical elements

Frequency Range - the ability to hear each element clearly and have all the frequencies of the audio spectrum properly represented

Panorama - placement of a musical element within the soundfield

Dimension - the addition of ambience to a musical element

Dynamics - control of the volume envelope of a track or instrument

and

Interest - making the mix special

It's possible to mix a song and only have four or five of these elements, but a great music mix contain all six, since they're all equally important (see Figure 4.2).

Figure 4.2: The 6 Elements Of A Mix

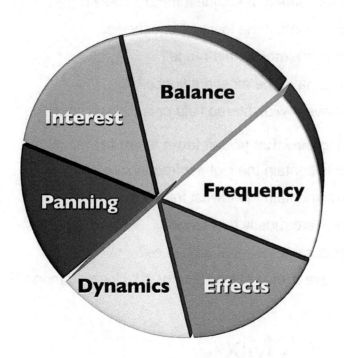

There are certain types of music that require the mixer to simply recreate an unaltered acoustic event like classical, jazz or a live concert recording, so sometimes only the first four elements are needed to have a mix be considered great. That being said, Dynamics and Interest have evolved to become extremely important elements as modern music has evolved.

► *Exercise 4.6*: *Identifying The Balance Elements Of A Mix*

A) **Listen to one of your favorite songs and listen to the balance of the mix.**

☐ What mix elements are in the forefront?

☐ What mix elements are in the background?

☐ How loud is the vocal compared to the other mix elements?

☐ Is the kick drum element louder than the bass element?

☐ Does the snare drum element drive the song?

B) **Now listen to a song at random and check out the balance of the mix.**

☐ What mix elements are the forefront?

☐ What mix elements are in the background?

☐ How loud is the vocal compared to the instruments?

☐ Is the kick drum element louder than the bass?

☐ Does the snare drum drive the song?

C) Listen to a song from a genre that you're unfamiliar with.

☐ Listen to the balance of the mix.

☐ What mix elements are the forefront?

☐ What mix elements are in the background?

☐ How loud is the vocal compared to the instruments?

☐ Is the kick drum element louder than the bass?

☐ Does the snare drum element drive the song?

▶ Exercise 4.7: *Identifying The Frequency Elements Of A Mix*

A) Go back and listen to that same favorite song and listen closely to the frequency balance of the mix.

☐ What mix elements have lots of high end?

☐ What mix elements are mid-range heavy?

☐ What mix elements have a lot of low end?

☐ What is the tone of the vocal?

☐ Are all mix element defined and clearly heard?

B) Now listen to that song you picked at random. Listen to the frequency response of the mix.

☐ What mix elements have lots of high end?

☐ What mix elements are mid-range heavy?

☐ What mix elements have a lot of low end?

☐ What is the tone of the vocal? Bright or mellow?

☐ Are all the mix elements defined and clearly heard?

C) Listen to that song from a genre that you're unfamiliar with. Listen to the frequency response of the mix.

☐ What mix elements have lots of high end?

☐ What mix elements are mid-range heavy?

☐ What mix elements have a lot of low end?

☐ What is the tone of the vocal?

☐ Are all the mix elements defined and clearly heard?

▶ Exercise 4.8: Identifying The Panning Elements Of A Mix

A) Go back and listen to that same favorite song.

☐ Where are all the mix elements placed in the stereo field?

☐ What mix elements are placed in the center?

☐ Is there a mix element that's panned to one side with an effect panned to the other?

☐ Are there mix elements that are doubled and panned to either side?

☐ Are there mix elements panned all the way to the left or right?

B) Now listen to that song you picked at random.

☐ Where are all the mix elements placed in the stereo field?

☐ What mix elements are placed in the center?

☐ Is there a mix element that's panned to one side with an effect panned to the other?

☐ Are there mix elements that are doubled and panned to either side?

☐ Are there mix elements panned all the way to the left or right?

C) Listen to that song from a genre that you're unfamiliar with.

☐ Where are all the mix elements placed in the stereo field?

☐ What mix elements are placed in the center?

☐ Is there a mix element that's panned to one side with an effect panned to the other?

☐ Are there mix elements that are doubled and panned to either side?

☐ Are there mix elements panned all the way to the left or right?

▶ Exercise 4.9: Identifying The Effects Elements Of A Mix

A) Go back and listen to that same favorite song.

☐ What mix elements or vocals seem dry and in your face?

☐ What mix elements or vocals can you hear reverb on?

☐ What mix elements or vocals can you hear delayed?

☐ Is the delay repeating?

☐ Does the snare drum element have an effect on it?

☐ Does the vocal have an effect on it?

☐ Can you hear any chorus or flanging on any of the mix elements?

B) Now listen to that song you picked at random.

☐ What mix elements or vocals seem dry and in your face?

☐ What mix elements or vocals can you hear reverb on?

☐ What mix elements or vocals can you hear delayed?

☐ Is the delay repeating? Does the snare drum have an effect on it?

☐ Does the vocal have an effect on it?

☐ Can you hear any chorus or flanging on any of the instruments?

C) Listen to that song from a genre that you're unfamiliar with.

☐ What mix elements or vocals seem dry and in your face?

☐ What mix elements or vocals can you hear reverb on?

☐ What mix elements or vocals can you hear delayed?

☐ Is the delay repeating?

☐ Does the snare drum element have an effect on it?

☐ Does the vocal have an effect on it?

☐ Can you hear any chorus or flanging on any of the mix elements?

> **Exercise 4.10:** *Identifying The Dynamics Elements Of A Mix*

A) Go back and listen to that same favorite song.

☐ Does the mix breathe in volume or does it seem compressed?

☐ Does the vocal sound compressed or natural?

☐ Do the kick and snare elements sound compressed or natural?

☐ Are there any mix elements that sound overly compressed?

☐ Is there a mix element that sounds like it needs more compression?

B) Now listen to that song you picked at random.

☐ Does the mix breathe in volume or does it seem compressed?

☐ Does the vocal sound compressed or natural?

☐ Do the kick and snare elements sound compressed or natural?

☐ Are there any mix elements that sound overly compressed?

☐ Is there a mix element that sounds like it needs more compression?

C) Listen to that song from a genre that you're unfamiliar with.

☐ Does the mix breathe in volume or does it seem compressed?

☐ Does the vocal sound compressed or natural?

☐ Do the kick and snare elements sound compressed or natural?

☐ Are there any mix elements that sound overly compressed?

☐ Is there a mix element that sounds like it needs more compression?

▶ E4.11: Identifying The Interest Elements Of A Mix

A) Go back and listen to that same favorite song.

☐ What's the most interesting mix element in the song?

☐ Is there a hook that grabs you?

☐ Is it a vocal or instrumental hook?

☐ Is it an effect?

B) Now listen to that song you picked at random.

☐ What's the most interesting mix elements in the song?

☐ Is there a hook that grabs you?

☐ Is it a vocal or instrumental hook?

☐ Is it an effect?

C) Listen to that song from a genre that you're unfamiliar with.

☐ What's the most interesting mix elements in the song?

☐ Is there a hook that grabs you?

☐ Is it a vocal or instrumental hook?

☐ Is it an effect?

Additional Ear Training Exercises

Now is a good time to check out the additional listening training at *Pro Audio Ears*. This is an excellent platform to train your ears not only how to listen, but what to listen for as well. It can take years off the time required to gain the experience needed to hear through a mix with confidence and authority. Go to ProAudioEars.com to register.

The Parts Of A Musical Arrangement

You read about the six elements of a mix in the last chapter, now let's look at the five parts of an arrangement. The arrangement of the song is so important because that's were the balance really begins. In a well-arranged song, you'll frequently find that it will almost mix itself. If the arrangement is, shall we say "less than adequate," you can fight and fight to fit everything in to make all the mix elements seem like they belong together.

That's why it's important to not only balance what you hear, but control what you don't hear as well. That means that sometimes muting a track is just as important to the mix as moving the faders.

So how does that relate to the arrangement? The number of musical elements that occur at the same time is critical to the attention of the listener. If there are too many playing at the same time, the listener gets confused, fatigued, and moves on. *That's why it's important for the mixer to have no more than five arrangement elements happening at the same time.* These elements are:

> **The Foundation** - The foundation is whatever mix element that provides the groove of the song. It's usually the bass and drums, but can also include a rhythm guitar and/or keys if they're playing the same rhythmic figure as the rhythm section. In electronic music it can mean a percussion or drum loop, or a synthesizer track.
>
> **The Pad** - A pad is a long sustaining note or chord. Organs, electric pianos, synthesizers, strings and even guitar power chords are the typical instrument pads that you'll find in a mix.
>
> **The Rhythm** - The rhythm is any mix element that adds motion and excitement to the track. A double time shaker or tambourine, a rhythm guitar strumming on the backbeat, or congas playing a Latin feel are all common example of a rhythm element.
>
> **The Lead** - A lead vocal, lead instrument, or solo.
>
> **The Fills** - Fills generally occur in the spaces between Lead lines, or answer the Lead. A common fill element can also be a signature line in an intro or interlude.

Most of the time, a song has only four or even three arrangement elements occurring at the same time, with the fills or the pad most likely to be left out.

You might think that about all the songs these days have a lot more than five tracks (some have as many as a hundred or more), so how does this work in the real world?

An element can be a single instrument like a lead guitar, synth loop, or a vocal, or it can be a group of instruments like the bass and drums, a doubled guitar line, a group of backing vocals, a loop from a song, etc. ***Generally, a group of instruments each playing exactly the same rhythm is considered an element.*** For example, a doubled lead guitar is a single element, as is a doubled lead vocal, or a lead vocal with two additional harmony vocals. Two lead guitars playing different parts are two elements, however. A lead and a rhythm guitar are two separate elements as well (see Figure 4.1).

In the song "It's About Time" that accompanies this book you can think of the horns (8 tracks of them) as just one element. Likewise, all the harmony and background vocals are just a single element. And if you go even further, all of the keyboards are playing the same basic rhythm, so they can be considered a single mix element as well, as are the acoustic guitars.

When you first look a mix with a lot of tracks it can seem intimidating at first, but once you really hear what they're playing you discover that there really are still usually only 5 arrangement elements or less that are playing at one time.

Figure 4.3: The Arrangement Elements

ELEMENT	PURPOSE	TYPICAL INSTRUMENTS
Foundation	The instruments that provide the groove or pulse of the song	Drums and bass
Pad	Long sustaining notes that glue the mix elements together	Organ, electric piano, strings, guitar power chords
Rhythm	The instruments that provide motion to the song	Percussion, rhythm guitar
Lead	The focal point of the song	Lead vocal, lead or solo instrument
Fill	The instruments that fill in the spaces between the lead phrases	Solo instrument, background vocal

Just so you can understand the arrangement elements of a song, we're going to look at a couple of recent huge hits as well as a rock standard from the 70s as examples.

"Uptown Funk" - Mark Ronson with Bruno Mars

- **The Foundation** - This one's easy, since it's the pretty standard bass and drums.

- **The Rhythm** - The rhythm element is an instrument that pushes the track along, usually double time or a complimentary rhythm to the foundation. In this song it's both the low "doh, doh, da" and the rhythm guitars.

- **The Pad** - A big synth sound during the chorus.

- **The Lead** - As almost always, it's the lead vocal.

- **The Fills** - The horn fills and background vocal answers.

"Girl Like You" - Maroon 5

- *The Foundation* - The kick, the claps and the bass.

- **The Pad** - It's a synthesizer that you can hear predominantly in the first verse, but it's there for the entire song adding the glue to the mix as well.

- **The Rhythm** - A filtered picking guitar that's low during most of the mix but adds all the motion.

- **The Lead** - As almost always, it's the lead vocal.

- **The Fills** - Synth and vocal answers in the spaces, especially during the second verse.

"Refugee" - Tom Petty And The Heartbreakers

- *The Foundation:* As with most songs, the foundation element for "Refugee" is held down by the drums and bass.

- *The Pad:* You can't get a better pad element than a Hammond B-3, and that's what you hear here.

- *The Rhythm:* There's a shaker that's placed low in the mix so it's not obvious, but it really pushes the song along with a double-time feel and you'd miss it if it wasn't there.

- *The Lead:* Tom Petty's lead vocal, and Mike Campbell's tasty guitar in the intro and solo.

- *The Fills:* Once again it's the guitar in the verses and the background vocal answers in the chorus.

Now let's look at the example tracks that go along with this book.

"It's About Time" - Billy White Jr.

- *The Foundation:* The drums and bass.

- *The Pad:* The 3 keyboard tracks - Wurlitzer, Rhodes and Piano (notice how they're layered in different octaves so they all fit together). Sometimes the Rhodes plays a fill as well.

- *The Rhythm:* There's lots of motion in this song with 3 percussion tracks - bongos, Tamborine and shaker. During the verse the sitar also acts as a rhythm element.

- *The Lead:* The lead vocal and the solo guitar in the intro, interlude and solo.

- *The Fills:* Background vocals and horns.

"Hard Plastic" - Bobby Owsinski

- *The Foundation:* The kick, claps, hat and bass.

- *The Pad:* This is an easy one - Pad 1.

- *The Rhythm:* Bongos and hat

- *The Lead:* FM Synth.

- *The Fills*: None, although the Reverse Cymbal might be thought of as a fill.

▶ Exercise 4.12: *Identifying Arrangement Elements*

A) Listen to your favorite song. Can you identify:

☐ the foundation?

☐ the pad?

☐ the rhythm?

☐ the lead?

☐ the fills?

B) Listen to a song at random. Can you identify:

☐ the foundation?

☐ the pad?

☐ the rhythm?

☐ the lead?

☐ the fills?

C) Listen to the song you're about to mix. Can you identify:

☐ the foundation?

☐ the pad?

☐ the rhythm?

☐ the lead?

☐ the fills?

◆

Chapter 5
Balance

The essence of mixing is the balance between instruments, virtual instruments, samples, loops, and vocals (which we'll call mix elements). No matter how good you are at other aspects of the mixing process, if you don't get the balance right, you don't have a mix. Over the years I mixed by ear like everyone else, with hit and miss results. Some of the old timers gave me some advice about how they did it, and it helped me to come up with a balance shortcut by the meters for balance, which I'm going to show you.

Building The Mix

There are actually a couple of directions to go when it comes to building a mix and both are valid. The first way is building the mix by ear, which can be somewhat unrepeatable when it comes to balances. Sometimes you get it just right, and other times you struggle to find the right combination of levels. Obviously the more experience you have the better and more consistent you are at it. The second method is designed to get you in the ballpark a lot faster and that's building a mix by using the meters. Let's look at both.

The Mix By Ear Method

When building a mix by ear, there is no standard mix element to start and build a mix from, although most mixers start with the kick. Modern mixers employ various approaches and they all work, especially in different genres of music. For instance, here are the places from which a mix by ear can be started:

- From the Bass
- From a loop
- From the Kick Drum
- From the Snare Drum
- From the Drum Overheads
- From the Lead Vocal or main instrument
- With all of the instruments and vocals in the mix right from the beginning
- When mixing a string section, from the highest string (violin) to the lowest (bass)

There are some mixers that just push up all the faders and start to mix with everything in the mix, which is similar to the second way that we'll build a mix.

The theory here is that everything will eventually be in the mix anyway, so you might as well start with all the mix elements there as soon as you can. The advantage to this method is that by hearing all the instruments, loops, and vocals, you're able to make an aural space in the mix for everything. If you insert one mix element at a time and tweak it so it sounds great, you begin to run out of space for the lead element (usually the vocal) so it never sits at the right level in the track. When that happens, you might have to go back to the beginning to make sure everything fits together properly.

Wherever you start from, it's a good idea that the lead arrangement element (usually the the vocal) be inserted into the mix as soon as possible. Since the vocal is the most important element, it may use up more frequency space than other supporting instruments. Many mixers find that it's difficult to find the right level with the rest of the track by waiting until late in the mix to insert the vocal.

The Mix By Meters Method

The second approach to building a mix is by setting each mix element to a certain level on the master buss meter. This technique was actually created back in the 1960s when mixers only had VU meters on consoles to reference to, but was refined later on when mixes for multiple musical artists during live radio or TV broadcasts had to be created fast.

TIP: Mixing by meters is not designed to be the final mix, but will get your mix in the ball park quickly. Remember that you most likely will have to tweak the balances a fair amount if compression and EQ are applied later.

To use this method first go to the loudest part of the song, then *watch only the master buss meters (NOTE: In Logic Pro be sure that both the Stereo Out and Master faders are set at 0, meaning the white line on the fader scale).*

- Start with the Kick track. Set the level so that only the peaks are touching -5dB on the master meter. If there are multiple Kick tracks, then all tracks played together should touch -5dB.

- Set the Snare track so that only the peaks are touching -5dB on the master meter. If there are multiple Snare tracks, then all tracks together should touch -5dB.

- Set the Tom tracks so only the peaks are touching -5dB on the master meter. This means the Tom hits only, not the leakage in between.

- Set the High Hat track so only the peaks are touching -20dB on the master meter.

- Set the Overhead or Cymbal tracks so only the peaks are touching -20dB on the master meter. This means the cymbal hits only, not the leakage in between.

- Set the Room track(s), if any, so only the peaks are touching -30dB on the master meter.

- Set the Percussion tracks so only the peaks are touching -25dB on the master meter. If there are multiple Percussion tracks, then all tracks together should touch -25dB. *Note: low frequency percussion like bongos or congas sometimes work better at -20dB.*

- Set the Bass track(s) so that only the peaks are touching -10dB on the master meter. If there are multiple tracks, then all tracks together should touch -10dB. *Note: If the Bass track has a high degree of low end content, then -15dB might work better instead.*

- Set the Vocal track so only the peaks are touching -10dB on the master meter. *Note: In some genres of music like rock or dance, the vocal might work better at -15dB instead.*

- Set the Guitar tracks so that only the peaks are touching -20dB on the master meter. If there are multiple tracks playing together at the same time, then all tracks together should touch -20dB.

- Set the Keyboard tracks so that only the peaks are touching -20dB on the master meter. If there are multiple tracks playing together at the same time, then all tracks together should touch -20dB.

- Set the Background Vocal tracks so that only the peaks in the song are touching -15dB on the master meter. If there are multiple tracks, then all tracks together should touch -15dB.

- Set the Vocal Double track (if there is one) so that only the peaks are touching -15dB on the master meter.

- Set the Horn tracks or Fill tracks so that only the peaks are touching -15dB on the master meter. If there are multiple tracks playing together at the same time, then all tracks together should touch -15dB.

- Set the Solo track so that only the peaks are touching -15dB on the master meter.

TIP: *The more precisely you adjust these levels, the more likely the mix will sound better when you're finished.*

To summarize the method on a chart, here's what the meter levels would look like, complete with all the variables.

Figure 5.1: Building A Mix Via The Meter Method

TRACK	MASTER METER LEVEL	VARIABLES
Kick	-5dB	
Snare	-5dB	
Toms	-5dB	Only when played; not leakage
High Hat	-20dB	
Overheads (Cymbals)	-20dB	Only when played; not leakage
Room	-30dB	increase for extra ambiance
Percussion	-25dB	Try low frequency percussion like bongos or congas at -20dB
Bass	-10dB	-15dB if extra bass content
Vocal	-10dB	-15dB for rock, dance or similar genres
Background Vocals	-15dB	All background vocal tracks together
Doubled Vocal	-15dB	Set 5dB less than lead vocal
Guitars	-20dB	All guitars tracks together
Keys	-20dB	All keyboards tracks together
Horns	-15dB	All horn tracks together
Claps	-15dB	All clap tracks together
Solo	-15dB	Set to -10dB if too low in the mix

TIP: The Meter Method only works with Sample Peak meters, which are usually the default setting of most DAWs. If the meters are set to any other ballistic measurement, then the mix will be unbalanced.

The Drums

In the early days of recording there was no such thing as balancing the drums since the entire kit was treated as a single instrument and miked with just a single mic. As producers began to understand how important the beat was, a mic was added to the kick. Eventually the modern drum sound evolved to where each drum and sometimes each cymbal is individually miked. As a result, the internal mix of the drums is a very important part of virtually every modern recording, unless you're using song loops where this balance is already set.

When mixing by ear, different engineers approach the drum mix in different ways. Some begin with the kick drum and build around that, while others start with the snare, since it provides the backbeat of most songs. Yet others want to build their drum mix around the toms so they don't get lost in the mix, especially if they're prominently featured.

A unique case has the mix being built around the overhead mics. The overhead mics are placed further away from the cymbals than normal cymbal miking because they're meant to pickup the overall sound of the drum kit. If overheads are used, many mixers like to start their mix from there and then fill in the sound with the other drum mics. This won't work so well when the mics are placed lower with the idea of just picking up the cymbals.

Remember that if you're dealing with real drums, the sound of every drum will change anywhere from a little to a lot when a new drum or cymbal is added to the mix due to the leakage from the other drums into the mic.

TIP: Microphone leakage is normal, and although there are ways to attenuate and even eliminate it, sometimes the overall drum sound suffers as a result.

Setting The Right Levels

In the Meter Method we start with the kick drum at -5dB. This is because we are mixing to very precise points on the master buss meter and if those levels are observed, the overload indicators on the master buss shouldn't light despite the high level.

In the By Ear method we start with the kick at -10dB. That's because you're more likely to make one or more tracks louder than what might be ultimately needed and cause an overload to occur, so we start with a lower level to compensate.

▶ *Exercise 5.1: Balancing A Song (Meter Method)*

Let's start with the "Hard Plastic" example song. This contains only loops of virtual instruments, but it's only 10 tracks. Of course, you can use your own track as well and still follow the exercise.

A) Raise the level of the Kick so it reads -5dB on the master meter.

B) Solo the Claps channel. Raise it's level until it reaches about -5dB on the master meter. Unsolo the channel so you can hear the full mix.

☐ Did the sound of the kick channel change when it was paired with the snare?

☐ Is the kick element masked by the snare and no longer distinct?

☐ How high does the master mix buss meter read?

C) Solo the High Hat channel. Raise the level of the channel until the master meter reaches -20dB. Unsolo the channel so you can hear the full mix.

☐ Does the sound of the snare element change?

☐ Does the sound of any of the toms or cymbals change?

☐ How high does the master mix buss meter read?

D) Solo the Bass channel. Raise the level of the channel until the master meter reaches -10dB. Unsolo the channel so you can hear the full mix.

☐ Does the kick and bass sound like they're at about the same level? If the bass sounds too loud, lower it to -15dB.

☐ What does the master mix buss meter read?

☐ Does the bass element sit well with the drums?

☐ Can you hear the kick, claps and bass elements distinctly? Don't worry if you can't. We'll fix it later with the EQ. Does the sound of any of the toms or cymbals change?

☐ How high does the master mix buss meter read?

E) Now solo the Bongo channel and set it so the peaks hit at -20dB on the master buss meter. Unsolo the channel so you can hear the full mix.

☐ Does it sound louder or quieter than you thought it would?

☐ Is it emphasizing a certain beat?

☐ How high does the master mix buss meter read?

F) Solo the Active Pad channel and set to -20dB on the master meter. Unsolo the channel so you can hear the full mix.

☐ Does it still stick out?

☐ Does it blend with the other mix elements?

☐ How high does the master mix buss meter read?

G) Solo the Bright Digi channel and set to -20dB on the master meter. Unsolo the channel so you can hear the full mix.

☐ Does it still stick out?

☐ Does it blend with the other mix elements?

☐ How high does the master mix buss meter read?

H) Solo the Pad channel and set to -20dB on the master meter. Unsolo the channel so you can hear the full mix.

☐ Does it still stick out?

☐ Does it blend with the other mix elements?

☐ How high does the master mix buss meter read?

I) Solo the FM Synth lead element and set to -10dB on the master meter. Unsolo the channel so you can hear the full mix.

☐ Does it stick out of the track or blend in with the other instruments?

☐ How high does the master mix buss meter read?

J) Go to the place in the song where the Reverse Cymbal plays and solo it. Set to -15dB on the master meter. Unsolo the channel so you can hear the full mix.

☐ Does it stick out of the track or blend in with the other instruments?

☐ How high does the master mix buss meter read?

_____

▶ Exercise 5.2: Building The Drum Mix From The Kick (Meter Method)

Now let's go to the "It's About time" example song. This song is considerably more sophisticated, with 52 tracks of real instruments and vocals. Of course, you can use your own track as well and still follow the exercise.

A) Raise the level of the kick drum channel until it reads about -5dB on the master mix bus meter.

B) Solo the Snare or Snare Top channel (if you're mixing the "It's About Time" example). Raise the level of the Snare Top channel until it's about the -5dB on the master meter. Unsolo the channel so you can hear the full mix.

☐ Did the sound of the kick channel change when it was paired with the snare?

☐ Is the kick element masked by the snare and no longer distinct?

☐ How high does the master mix buss meter read?

C) Solo the Snare Bottom channel (if you're mixing the "It's About Time" example) along with the Snare Top. Raise the level of the Snare Bottom channel until the Snare sound begins to brighten. Then make sure that both Snare channels read about -5dB on the master meter. Unsolo the channels so you can hear them in the full mix.

☐ Did the sound of the overall snare element change when it was paired with the snare bottom?

☐ Did the sound of the kick element change when it was paired with the snare?

☐ Is the kick element masked by the snare and no longer distinct?

☐ How high does the master mix buss meter read?

D) Go to a place in the song where there are Tom fills (look for the hits on the timeline). Solo the Tom channels. Raise the level of all Tom channels until they hit -5dB on the master meter. Unsolo the channels so you can hear the full mix.

☐ Did the sound of the kick element and/or snare change?

☐ Do the kick and snare elements sound different when the toms aren't playing?

☐ How high does the master mix buss meter read?

E) Solo the Cymbal or Overhead channels. Raise the level of the channels until it hits around -20dB on the master meter. Unsolo the channels so you can hear the full mix.

☐ What happened to the sound of the other drums?

☐ Is the drum sound fuller or does it contain more ambience?

☐ How high does the master mix buss meter read?

F) Solo the High Hat channel. Raise the level of the channel until the master meter reaches -20dB. Unsolo the channel so you can hear the full mix.

☐ Does the sound of the snare element change?

☐ Does the sound of any of the toms or cymbals change?

☐ How high does the master mix buss meter read?

G) Solo the Room channel(s). Raise the level of the channels (if you're mixing the "It's About Time" example) until the master meters reach -30dB. Unsolo the channel so you can hear the full mix.

☐ What happened to the sound of the other drums?

☐ Does the drum sound have more ambience?

☐ How high does the master mix buss meter read?

> **Exercise 5.3:** *Building From Drum Mix From The Kick (By Ear Method)*

A) **Raise the level of the kick drum channel until it reads about -10dB on the master mix bus meter.**

B) **Solo the Snare Top channel (if you're mixing the "It's About Time" example). Raise the level of the Snare Top channel until it's about the same level on the meter. Unsolo the channel so you can hear the full mix.**

☐ Did the sound of the kick channel change when it was paired with the snare?

☐ Is the kick element masked by the snare and no longer distinct?

☐ How high does the master mix buss meter read?

C) **Solo the Snare Bottom channel (if you're mixing the "It's About Time" example). Raise the level of the Snare Bottom channel until it's about the same level on the meter. Unsolo the channel so you can hear the full mix.**

☐ Did the sound of the snare element change when it was paired with the snare bottom?

☐ Did the sound of the kick element change when it was paired with the snare?

☐ Is the kick element masked by the snare and no longer distinct?

☐ Is the snare sound now too bright or unnatural. If so, back off the fader of the Snare Bottom channel until you like the sound.

☐ How high does the master mix buss meter read?

D) **Go to a place in the song where there are Tom fills (look for the hits on the timeline). Solo the Tom channels. Raise the level of all Tom channels until they're about the same level on the meter as the kick and snare. Unsolo the channel so you can hear the full mix.**

☐ Did the sound of the kick element and/or snare change?

☐ Do the kick and snare elements sound different when the toms aren't playing?

☐ How high does the master mix buss meter read?

E) **Raise the level of the Cymbal or Overhead channels until the overall sound begins to change and the cymbals become more distinct sounding.**

☐ What happened to the sound of the other drums?

☐ How high does the master mix buss meter read?

F) Raise the level of the high-hat channel until it becomes a bit more distinct sounding.

☐ Does the sound of the snare element change?

☐ Does the sound of any of the toms or cymbals change?

☐ How high does the master mix buss meter read?

G) Raise the level of the Drum Room channels (if you're mixing the "It's About Time" example) until the overall drum sound begins to change.

☐ What happened to the sound of the other drums?

☐ Does the drum sound have more ambience?

☐ How high does the master mix buss meter read?

Checking The Drum Phase

One of the most important yet overlooked parts of a drum mix is checking the phase of the drums. This is important because not only will an out-of-phase channel suck the low end out of a mix, but it will get more difficult to fix as the mix progresses.

A drum mic can be out of phase due to a mis-wired cable or poor mic placement. Either way, it's best to fix it now before the mix goes any further.

TIP: The phase switch or selector is usually located at the top of the module near the input section on a hardware mixer or console, or in various locations of most plugins (see Figure 5.2).

▶ *Exercise 5.4: Checking The Drum Phase*

If you're not using a hardware console or mixing with a phase selection switch built into each channel strip, insert a plugin that has phase selection on each drum channel for the following exercise.

A) With all the drums in a balanced mix, solo the Kick drum channel.

B) Now solo the Snare channel and change the selection of the polarity or phase control (see Figure 5.2).

☐ Is there more low end or less? Chose the selection with the most bottom end.

C) Solo each Tom channel by itself (just Tom and Kick) and change the selection of the polarity or phase control.

☐ Is there more of less bottom end? Chose the selection with the most bottom end.

D) Solo each Cymbal or Overhead channel (just Overhead and Kick) and change the selection of the polarity or phase control.

☐ Is there more low end or less? Chose the selection with the most bottom end.

E) If there is a Drum Room channel (like in the "It's About Time" example), solo the channel and change the selection of the polarity or phase control.

☐ Is there more low end or less? Chose the selection with the most bottom end.

F) If there is a Snare Bottom channel (like in the "It's About Time" example), solo both the Snare Top and Snare Bottom channels only (unsolo the Kick) and change the selection of the polarity or phase control on the Snare Bottom channel.

☐ Is there more low end or less? Chose the selection with the most bottom end.

Figure 5.2: A Channel Phase Control (inside the box with the arrow pointing to it)

Assigning The Drums To A Group Or Subgroup

Whenever there are two or more channels that make up a mix element, like a drum kit, three guitars, six background vocals, or eight percussion tracks (for example), it's best to assign them to either a group or a subgroup in order to make any mix adjustments that you might have to make later somewhat easier.

TIP: In a group, a number of channel faders (like the drums) are electronically or digitally linked together so that if you move one fader in the group, they all move, yet keep the same relative balance that you originally set.

A subgroup has the same effect as a group yet works a little differently. All of the channels of the group are assigned to a subgroup fader, which is then assigned to the master mix buss. The level of all the channels is controlled with that one subgroup fader, and if you move any fader within the group, the others don't move with it but you change the balance of the submix. If you send to an effect or insert a compressor from the subgroup it also affects all the instruments in that group, which can be a benefit in certain situations during mixing.

▶ *Exercise 5.5: Creating a Drum Group Or Subgroup*

In order to perform the two parts of this exercise you might have to refer to your console or DAW application manual.

A) On your DAW, assign all of the drum channels to a group. Move one of the faders.

   ☐ Does the sound of the snare element change?

   ☐ Does the sound of any of the toms or cymbals change?

   ☐ How high does the master mix buss meter read?

B) On your console or DAW, assign all of your drum channels to a subgroup. Raise or lower the subgroup fader.

   ☐ Does the level of the drums get louder and softer as you move the fader?

   ☐ Are all the channels assigned to the subgroup silenced when the subgroup mute is engaged?

The Percussion Element

Percussion is divided into two classes - the high-frequency type, which includes shakers, maracas and tambourines; and the low-frequency type that includes congas and bongos. Percussion has one purpose and that's to add motion to a song, so therefore it's a Rhythm element in the mix.

Because percussion generally has a very fast transient response with a lot of peaks, it doesn't take much level for it to stick out of a mix. Mix high-frequency percussion at -25dB on the master buss meter. Low-frequency percussion might work at the level as well, but usually -20dB is better.

► Exercise 5.6: Mixing Percussion

This exercise refers to the "It's About Time" song example.

A) Solo the Tambourine channel and set it to -25dB. Then unsolo and listen to the mix.

☐ Does it sound louder or quieter than you thought it would?

☐ Is it emphasizing a certain beat of the song?

B) Unsolo the Tambourine channel and solo the shaker channel. Set it to -25dB, then unsolo and listen to the mix.

☐ Does it sound louder or quieter than you thought it would?

☐ Is it emphasizing a certain beat?

D) Now solo both the Tambourine channel and the shaker channels. Make sure both now hit -25dB on the master buss meter.

☐ Can you hear both clearly or is one covering the other?

E) Unsolo both the Tambourine channel and the shaker channels so you hear the entire mix.

☐ Can you hear the percussion clearly in the mix?

☐ How high does the master mix buss meter read?

F) Now solo the Bongo channel and set it so the peaks hit at -20dB on the master buss meter.

☐ Does it sound louder or quieter than you thought it would?

☐ Is it emphasizing a certain beat?

G) Unsolo the Bongo channel so you hear the entire mix.

☐ Can you hear the bongos clearly in the mix?

☐ How high does the master mix buss meter read?

H) Now go to the place in the song with the shaker. Solo the Shaker channel and set it so the peaks hit at -25dB on the master buss meter.

☐ Does it sound louder or quieter than you thought it would?

☐ Is it emphasizing a certain beat?

I) Unsolo the Shaker channel so you hear the entire mix.

☐ Can you hear the shaker clearly in the mix?

☐ How high does the master mix buss meter read?

The Bass Element

The balance between the bass element and drums is critical because that's where the power of the mix comes from. Where once upon a time the bass amp of a live bass was always miked, today most basses are taken direct, but sometimes both the amp mic and the direct signals are recorded on separate tracks as well. This technique might be used so that the bass sound has the best combination of bottom end and clarity. Of course, sometimes the bass element is created via a MIDI sample or virtual instrument and might be layered with multiple tracks as well.

Just like the drums, the phase between the direct bass sound and the miked one must be checked if both have been recorded because if they're out of phase, the bass may sound thin with no power. The same is true if you're using multiple samples or virtual bass instruments in an effort to get a bigger sound.

TIP: Sometimes a bass element that sounds on the small side will sound a lot fuller when combined with the kick. That's why it's important to listen to the kick and bass elements together.

> ### Exercise 5.6: *Balancing The Bass Channels*

A) Pick the Bass channel you think sounds best and raise the level so it reads -10dB on the master mix buss meter.

B) Slowly raise the level of the second bass channel.
  - ☐ Does the bass sound fuller and fatter?
  - ☐ Does it sound clearer and more distinct?

C) Set it where you think it sounds best for now. This balance will be adjusted later when more mix elements are introduced into the mix.

D) Assign the bass channels to a group or subgroup for ease of level adjustment later.

E) Set the level of all bass channels together so they hit -10dB on the master meter.

> ### Exercise 5.7: *Checking The Bass Phase (if more than one track)*

First check the phase between the bass amp and direct signal if both channels were recorded.

A) Solo the bass channels again using the same balance from the previous exercise.

B) Change the selection of the polarity or phase control on the channel strip or plugin only on channel that's lowest in level.
  - ☐ Is there more low end or less? Chose the selection with the most bottom end.

▶ Exercise 5.8: *Balancing The Bass And Drum Elements (Meter Method)*

A) You've set the levels to the master meter in the previous exercises. Unmute or unsolo all tracks now so you can hear the bass and drum balance.

☐ Does the kick and bass sound like they're at about the same level? If the bass sounds too loud, lower it to -15dB.

☐ What does the master mix buss meter read?

☐ Does the bass element sit well with the drums?

☐ Can you hear the kick, snare and bass elements distinctly? Don't worry if you can't. We'll fix it later with the EQ.

TIP: Don't worry if the kick and snare sound particularly loud at this point. You'll find that the balance is close to what you need when all instruments are eventually inserted into the mix.

▶ Exercise 5.9: *Balancing The Bass And Drum Elements (By Ear Method)*

A) Using your final drum mix from before, mute all the drum channels except the kick. It should read about -10dB on the mix buss meter.

☐ Now mute the channel.

B) Raise the bass group or subgroup channel until the master mix buss meter reads -10dB.

☐ Now unmute the kick drum channel.

☐ Does the kick and bass sound like they're at about the same level?

☐ What does the master mix buss meter read?

C) Unmute the other drum channels so you hear the entire drum kit with the bass.

☐ Does the bass element sit well with the drums?

☐ Can you hear the kick, snare and bass elements distinctly? Don't worry if you can't. We'll fix it later with the EQ.

The Vocals

Many mixers like to get the vocal in the mix as soon as possible because if you wait until the end after you've mixed all the music, the vocal will probably be either too loud or too soft. Plus, the vocal is usually the focal point of the song, so it's best to get that sounding great and build around it.

TIP: in pop and country songs the lead vocal is often louder than the rest of the mix elements to keep it front and center. In a rock or dance mix, the vocal is usually lower than the other mix elements to emphasize the backing track and make it sound more powerful.

Since the 60s doubling a vocal has been an effective way to make it sound fuller and to even out any pitch inconsistencies. There are two ways to treat vocal doubling; set both vocals at the same level, or set one about 5 to 10dB lower than the lead vocal and use it for support.

Background Vocals

Most of the trick to background vocals has more to do with panning, equalization, and effects than it does with balance, but let's look at a couple of scenarios where balance does come into play.

Harmony Vocals - With harmony vocals, the balance is crucial in order to get the correct blend and impact. Usually the highest vocal cuts pretty well, but the lowest or one in the middle of a three part harmony gets lost. If the lowest part of the three part harmony is the melody, then it's usually the middle part that gets lost.

TIP: The easiest (but not the only) way to balance three part harmony is to begin just like you did with the rhythm section, from the bottom up. Start with the lowest vocal, add the middle vocal until the blend is such that they sound as one, then add the highest vocal part.

Make sure to assign the background vocal channels to either group or a subgroup.

Gang Vocals - Gang vocals are shouts or a unison vocal part where the balance isn't as important to attain a blend as with harmony vocals. If well recorded, the gang vocal will have a lot of different types of voices, and/or a lot of different natural room ambience because some parts may have been recorded further from the mic.

There are two ways to approach balance in this situation. If you have a lot of voices with different timbres, once again start from the lowest to the highest, since the highest sounding voices will cut through the mix easier. If you have voices with different room ambiences, start with the one that sounds furthest away to get the biggest sound.

> **Exercise 5.10**: *Balancing The Lead Vocals With The Rhythm Section (Meter Method)*

    A) Solo the lead Vocal and raise the fader until it reads -10dB on the mix buss meter.

    B) Unsolo the Vocal and listen to the vocal and bass and drums together.

        ☐ Do the drums seem to overwhelm the vocal?

        ☐ Does the vocal seem to overwhelm the drums?

        ☐ Can you hear every word of the vocal?

        ☐ What level does the master buss meters read?

> **Exercise 5.11**: *Balancing The Lead Vocals With The Rhythm Section (By Ear Method)*

    A) Raise the level of the lead vocal until it's about the same level as the rhythm section.

        ☐ Do the drums overwhelm the vocal?

        ☐ Does the vocal seem to overwhelm the drums?

        ☐ Can you hear every word of the vocal?

        ☐ What level does the master buss meters read?

    C) Continue adding mix elements around the rhythm section and lead vocal.

        ☐ What level does the master buss meters read when you add each mix element?

        ☐ Is the master buss overloading? If so, decrease the master fader.

> **Exercise 5.12:** *Balancing A Vocal Double*

    A) Go to a place in the song with the Vocal Double. After the vocal level is set as above, raise the level of the Double until it's the same level as the lead vocal.

        ☐ Does the vocal sound louder?

        ☐ Is the vocal fuller sounding?

    B) After the vocal level is set as above, raise the level of the double until it sits about -10dB below the lead vocal (or -20dB in this case), or loud enough that you can just hear it.

☐ Does the vocal sound louder?

☐ Is the vocal fuller sounding?

☐ Can you still hear any pitch problems?

C) Assign both vocals to a group or subgroup and readjust the level as in exercise E5.10 or E5.11.

▶ *Exercise 5.13: Balancing Background Vocals*

A) Meter Method - With the background vocals soloed but fully attenuated, raise the level of the lowest vocal to -15dB on the mix buss meter, then the middle vocal to the same level, then the highest vocal to the same level.

☐ Do they sound balanced in level?

☐ Do they blend so they sound like one voice? If not start again.

B) By Ear Method - Raise the level of the lowest vocal to -10dB on the mix buss meter, then raise the middle vocal level so that it sounds balanced, then the highest vocal so that it sounds balanced with the other two.

☐ Do they now sound balanced in level?

☐ Do they blend so they sound like one voice?

C) Assign the background vocal channels to a group or subgroup, unsolo, and set to -15dB.

▶ *Exercise 5.14: Balancing Gang Vocals*

A) With the gang vocal channels soloed but fully attenuated, raise the level of the lowest or deepest sounding vocal first, then add in all vocals in one by one, with the ones with the highest timbre last.

☐ Do they blend together so they sound like one voice?

☐ What does the master buss meters read?

B) With the gang vocals soloed, raise the level of the vocal that sounds furthest away first, then add in all vocals with the closest sounding vocal last.

☐ Do they blend together so they sound like one voice?

C) Assign the gang vocal channels to a group or subgroup, unsolo, and balance against the other mix elements.

Keyboards

Keyboards can act as just about any arrangement element. Their function in the mix is determined by the type of sound, whether it comes from a real instrument, virtual instrument, or synthesizer patch. These are piano, organ, electric piano, and synthesizers.

Depending upon what arrangement element the keyboard serves, it might be better to add it to the mix before any guitar tracks. This is up to you, but also depends upon the song and arrangement as well.

Piano

Since a grand or upright piano is very percussive in nature it can serve as either a foundation, rhythm, lead or fill arrangement element. It's possible that it can also be a pad as well if it plays long sustaining chords.

Electric Piano

The electric piano seems made for the pad element since it's very mellow sounding, easily blends in with the track, and is capable of playing long sustaining chords. You may find it used as other elements as well, but not nearly as often as the pad. Sometimes this is called Wurlitzer or Rhodes, the names of the real instruments the virtual instrument or sample emulates.

Organ

The organ is the perfect pad element since it's capable of infinite sustaining chords. While there are many types of organs, the Hammond organ sound in particular is often used as the glue to a track.

Synthesizers

Synths are another instrument that can also serve as any arrangement element. They can be very percussive and serve as a foundation element, or can create a great pad.

▶ Exercise 5.15: Balancing The Keyboards (Piano, Organ, Synthesizer)

A) Go to the section of the song where the piano plays and solo it.

☐ Does it belong to the rhythm section, is it a pad or rhythm element, or is it a lead or fill element?

B) Set the Piano channel so only the peaks hit -20dB on the master meter, then unsolo.

☐ Does it still stick out?

☐ Does it blend with the other mix elements?

C) Solo the Wurlitzer channel and set it so only the peaks hit -20dB on the master meter, then unsolo.

☐ Does it still stick out?

☐ Does it blend with the other mix elements?

D) Solo the Rhodes channel and set it so only the peaks hit -20dB on the master meter, then unsolo.

☐ Does it still stick out?

☐ Does it blend with the other mix elements?

E) Assign the Keyboard channels to a group or subgroup, solo, and set to -20dB. Unsolo and listen.

☐ Do the keyboards stick out of the track or blend in?

☐ Do the keyboards seem too loud or not loud enough? We can adjust this later.

E) If mixing by ear, mix the keyboard so it blends into the rhythm section.

☐ Does it still stick out?

☐ Does it blend with the other mix elements?

▶ *Exercise 5.16:* *Balancing A Keyboard That's A Lead Instrument*

A) Solo the lead element and set it to -10dB on the master meter. Unsolo and listen

☐ Does it stick out of the track or blend in with the other instruments?

☐ If too loud solo again and lower the level to -15dB.

Guitars

Just like keyboards, guitars can act as every type of arrangement element, depending upon the song. They can be part of the rhythm section by playing quarter notes with the snare drum; they can play big power chords that act as the pad of the song,; or they can strum in double time to push the song along as the rhythm element. If that isn't enough, they can be the lead instrument in solos, intros, or in an instrumental, or be fills, playing in the holes around the vocal.

Often multiple guitar tracks that play roughly the same part are record to achieve a bigger sound, which can cause problems that can't always be taken care of by balance alone. Some frequency adjustments of the tracks, which we'll cover in Chapter 7, are required to keep them from fighting each other.

By far, the primary task for the mixer is to identify exactly what arrangement element the guitar fits into, then balance it accordingly.

▶ *Exercise 5.17*: *Balancing The Electric Guitar*

A) Using the first song example, go to the intro of the song and solo Guitar 1.

☐ Does it belong to the rhythm section, is it a pad, or rhythm element, or is it a lead or fill element?

B) Set the level of the peaks of Guitar # so they just hit -20dB on the master meter. Unsolo the channel and listen.

☐ Does it still stick out or blend into the mix?

C) Repeat the procedure for every guitar in the mix except for the Solo. Remember, set only the peaks to -20dB on the master meter.

☐ Does it stick out of the track or blend in?

☐ Do the guitars seem too loud or not loud enough

C) Assign all the guitar channels to a group or subgroup, solo, and set to -20dB.

☐ Does the guitars stick out of the track or blend in?

☐ Do the guitars seem too loud or not loud enough

▶ *Exercise 5.18*: *Balancing The Acoustic Guitar*

A) Using the first song example, go to the part of the song where the Acoustic Guitars play and solo Acoustic 1. Solo it and listen.

☐ Does it belong to the rhythm section, is it a pad, or rhythm element, or is it a lead or fill element?

B) Set the level of the peaks of Acoustic 1 so it just hits -20dB on the master meter. Unsolo the channel and listen.

☐ Does it still stick out or blend into the mix?

C) Solo the Acoustic 2 channel and set it so it just hits -20dB on the master meter. Unsolo the channel and listen.

☐ Does it stick out of the track or blend in?

☐ Does Acoustic 2 sound different from Acoustic 1?

C) Assign the Acoustic Guitar channels to a group or subgroup, solo, and set to -20dB.

☐ Do the guitars stick out of the track or blend in?

☐ Do the guitars seem too loud or not loud enough? We can adjust this later.

▶ Exercise 5.19: *Balancing A Lead Guitar (Meter Method)*

Listen to the Solo Guitar in the example. It's a lead instrument in the intro, interlude and solo, and a fill element when used elsewhere.

A) Solo the Solo guitar and set it so the peaks hit -15dB on the master meter. Unsolo and listen.

☐ Does the Solo overpower the rhythm section or is it too low in the mix? Raise to -10dB if too low.

▶ Exercise 5.20: *Balancing A Lead Guitar (By Ear Method)*

A) Raise the level of the lead guitar until it's the loudest musical element.

☐ Does the vocal overpower the rhythm section?

B) Raise the level of the lead guitar so it's only as loud, or even a little softer, than the bass and drums.

☐ Does the track still have excitement when it plays?

▶ Exercise 5.21: *Balancing A Power Chord Guitar*

If mixing a rock or metal song with a distorted guitar playing power chords that are used as a pad:

A) Mix the guitar pad so it's down low in the track, just barely audible. Mute and unmute it.

☐ Does the track sound fuller when it's unmuted?

☐ Does it stick out?

B) Mix the guitar pad so it's up front, almost as loud as the vocal.

☐ Does it take attention away from the vocal?

☐ Is the song still powerful or does it lose the groove?

☐ Does the track still sound full if you mute it?

Horns

Horns are broken down into two categories - brass and woodwinds. Brass sections on a pop song will usually consist of trumpets and trombones, although french horns also pop up every now and then. Woodwinds consist of mainly saxophones (alto, tenor and baritone), flutes, clarinets, and in an orchestral setting oboes and bassoons.

When used as a solo instrument, any horn instrument is usually places louder in the mix (-15dB if you're using the meter method), but usually you'll be mixing them as a section (-20dB if lots of brass, -15dB if more presence in the mix is required).

▶ Exercise 5.22: *Balancing The Horn Section*

A) As with guitars and keyboards solo each Horn and make sure the peaks hit -20dB on the master meters.

B) Assign the Horn channels to a group or subgroup, solo, and set to -20dB.

☐ Do the horns stick out of the track or blend in?

☐ Do the horns seem too loud or not loud enough? We can adjust this later.

Loops And Samples

Loops are a crucial musical element these days and generally change the way you build a track, depending upon the importance of the element. If a loop is present, it's usually best to start the mix with that element, then build the mix around it, since the loop can be considered a foundation arrangement element or part of the rhythm section.

TIP: If the song is made up primarily of loops, then it's important to find the combination of loops that creates the groove, then build the mix from there.

▶ Exercise 5.23: *Balancing Beats, Loops, And Samples*

Many songs are augmented with additional beats, loops or samples and it's easy for them to get buried in the mix. Here are a couple of approaches to make them fit well.

A) Raise the level of the loop until you just begin to hear it in the track.

☐ Does it belong to the rhythm section?

☐ Is it a pad or rhythm element, or is it a lead or fill element?

☐ Does it fight other instruments for space?

B) For now, raise the level of the loop until it hits -20 on the master meter.

☐ Can it be heard in the mix? If no, raise to -15dB.

☐ Does it fight other instruments for space? If so, this is something that we'll fix later.

C) Begin the entire mix again only starting with the loop first. Raise it so the level on the mix buss meter reads -10dB, and add the other mix elements as in the previous exercises.

☐ Does the loop fit better into the mix now?

☐ Does the loop or the rhythm section provide the groove?

▶ *Exercise 5.24: Balancing A Song Based On Complex Loops*

A complex loop is one that already has most of the instruments mixed into it. This could be downloaded from an online service, sampled from an existing song (be sure to get written permission first), or provided by a beat maker. Additional overdubs and vocals will enhance the original loop or loops, but the basic structure of the song comes from the complex loop.

A) Raise the level of the loop until it hits -10dB on the master meter.

B) Add any overdubs or vocals into the mix at the same level as outlined in the previous exercises.

☐ Do the overdubs fight for space in the mix?

☐ Does the vocal fight for space in the mix? Don't worry, we will fix this later.

Mixing By Muting

It's not uncommon to work with an artist or band that isn't sure of the arrangement, or is into experimenting and just allows an instrument to play throughout the entire song. As a result, more than five arrangement elements are present, or some parts become inappropriate for the song in certain sections.

This is where the mixer gets a chance to rearrange the track by keeping what works and muting the conflicting instrument or instruments (see Figure 5.3). Not only can the mixer influence the arrangement this way, but also the dynamics and general development of the song as well. In fact, sometimes the track that's muted becomes more important that if it we left playing.

TIP: Always consider the mute buttons of the channels as a welcome feature that can help you blend your tracks together just as much as the faders.

Figure 5.3: The Channel Mute Button

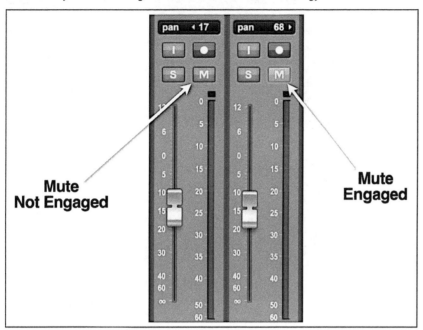

Exercise 6.25: *Final Balance*

Go to the outro of the song where all channels are playing.

A) Look at the balance between all the channels.

☐ What happens if you alter balance slightly so every channel is heard better?

☐ Are any tracks still masked?

☐ Can you hear every track distinctly? Don't worry if you can't. We'll fix that soon.

Chapter 6
Panning

One of the most taken-for-granted aspects in mixing pertains to the placement of a mix element in the stereo sound field, otherwise known as panning. Not only does stereo provide a sense of spaciousness, but it allows us to create excitement by adding movement within the sound field, as well as adding clarity to a mix element by moving it out of the way of others around it.

Moving a sound out of the way of another via the pan pot is really important in a mix. Sometimes just a little movement to the left and right will suddenly take a mix element from obscurity to definition. This works well with background and lead vocals, for example. When the lead and background vocals are panned in the center, each can obscure the other so that neither is heard very well. if the background vocals are moved a little out from the center, both can now be distinctly heard on their own.

THE TRICK TO PANNING: Pan mix elements to a vacant place in the sound field not occupied by another mix element.

The Three Main Panning Areas

There are three main areas of a sound field that are used the most; hard left, hard right, and center. When a sound of an instrument comes out of both speakers at an equally loud level, it seems as if it's coming from in between them. This is what's known as the "phantom center" (see Figure 6.1). This phantom center can shift from left to right as you move your head around, which is why the exact center between the speakers where the phantom image is heard is called "the sweet spot."

The other two main panning areas are hard to the left, and hard to the right, which means you place the pan pot all the way to the left or right so the so sound only is heard in that one speaker. This gives us 3 main panning areas that many engineers call "the big 3."

Panning a mix element hard left or hard right can be distracting, and panning to the center can get in the way of a lead mix element like a vocal, so you must have a good reason to choose any of these areas.

Figure 6.1: The Big 3

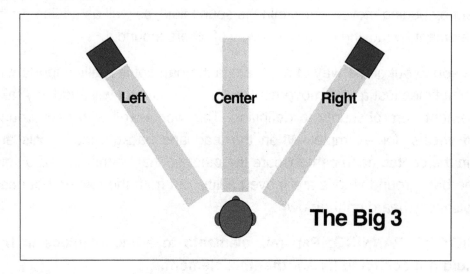

Big Mono

Big Mono occurs when you have a track with a lot of stereo mix elements that are all panned hard right and hard left. In this case, you're not creating much of a panorama because everything is placed hard left and right, and you're robbing the track of definition and depth because all of these tracks are panned on top of one another (see Figure 6.2).

The solution is to find a place to pan those mix elements somewhere inside those extremes. One possibility is to pan the left source to about 10:00 while the right is panned to about 4:00. Another more localized possibility would be to put the left to 9:00 and the right all the way to 10:30. This gives the feeling of localization without getting too wide.

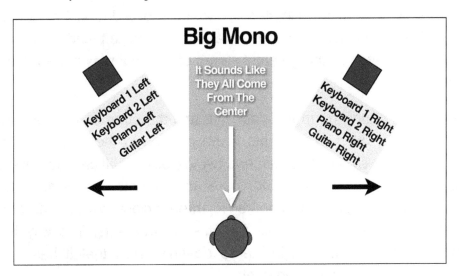

Low Frequencies In The Center

In most mixes the most prominent music element (usually the lead vocal or instrument) is panned in the center, but the kick, bass and even the snare mix elements can be found there as well. Putting the bass and kick in the middle makes the mix feel strong and anchored, but this practice really comes from the era of vinyl records.

In the early days of stereo back in the mid-sixties, the mixers of the day only had three-way pan switches that selected either the left, the right, or both speakers, which gave you the phantom center effect. The true pan pot that allowed you to place the instrument anywhere in between the speakers hadn't been invented yet. As a result, some of the early stereo records had the vocals on the left side and all the music on the right side, like on some of the early Beatle stereo releases.

The problem is that a mix with most of the low frequency information on one side caused a lot of problems when making the vinyl record, causing it to skip.

Today we don't have to worry about the limitations of the vinyl record anymore (unless that's your intended distribution medium), so it's possible to place the bass, kick drum, floor tom, or any instrument with a lot of low frequencies, anywhere in the stereo spectrum that we like, since it won't make a bit of difference to a music stream, AIFF file, or a CD. Still, mixers everywhere place anything with low frequency information in the center simply because it makes the mix sound much more powerful.

TIP: If you're using Logic Pro, right click on the pan control and select "Stereo Pan" for the panning of stereo tracks in the following exercises.

Panning Stereo Instruments

When an instrument is recorded in stereo, like a piano, a drum kit or doubled guitars, your first inclination is to pan everything hard left and hard right. Although that does give you the greatest stereo effect, that may not always be best for the mix.

Sometimes panning one side to 9 o'clock left, and the other at 1 o'clock right gives more room for other instruments to be panned to the right side (see Figure 6.3). Sometimes panning at ten and two o'clock is sufficient to hear the effect and keep some space open for other instruments. Sometimes even panning one side hard right and another to three o'clock will provide some stereo effect yet leave lots of space for other instruments. The key is to remember that just because it's in stereo doesn't mean that it has to be panned to the extreme hard left and right.

TIP: *Caution, panning all stereo mix elements hard left and hard right creates something known as "Big Mono," where the stereo effects are negated because there are so many elements on top of one another.*

Figure 6.3: A Stereo Instrument With A Narrow Pan

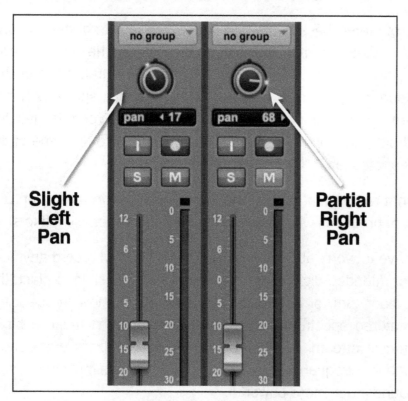

➤ *Exercise 6.1*: *Simple Panning*

Let's begin with the mix that you achieved for "Hard Plastic" in the last chapter.

A) Start by making sure that the panning of the Kick, Claps, Hat and Bass channels are panned to the center. Mute all other tracks.

B) Now umute the Bongos channel and pan it to the 11 and 1 o'clock positions.
- ☐ What happens to the sound?
- ☐ Does it still sound like stereo?
- ☐ What happens to the mix buss meters?

C) Unmute the Active Pad channel and pan it so that the left pan is at 10 o'clock and the right on is at 1 o'clock.
- ☐ Does the sound lean to one side?
- ☐ Does the Active Pad still sound like it's in stereo?
- ☐ What happens to the mix buss meters?

D) Unmute the Bright Digi channel and pan it so that the left pan is at 11 o'clock and the right on is at 2 o'clock.
- ☐ Does the sound lean to one side?
- ☐ Does the Active Pad still sound like it's in stereo?
- ☐ What happens to the mix buss meters?

E) Solo the Bright Digi and Active Pad channels and listen.
- ☐ Does the sound lean to one side?
- ☐ Does the mix still sound like it's in stereo?
- ☐ What happens to the mix buss meters?

F) Unsolo the Bright Digi and Active Pad channels, then unmute the Pad Channel and pan it to the 9 and 3 o'clock positions.
- ☐ Does the sound lean to one side?
- ☐ Does the mix still sound like it's in stereo?
- ☐ Can you hear all the instruments?
- ☐ What happens to the mix buss meters?

G) Unmute the FM Line and Reverse Cymbal channels and pan them to the center.
- ☐ Does the sound lean to one side?
- ☐ Does the mix still sound like it's in stereo?
- ☐ Can you hear all the instruments?
- ☐ What happens to the mix buss meters?

G) Go back and readjust the panning. Can you find better positions

_____

▶ *Exercise 6.2*: Big Mono

Let's stay with "Hard Plastic" for this exercise.

A) Solo the Bongo, Active Pad, Bright Digi and Pad channels.

B) Now pan each of those channels hard left and right.

☐ Can you hear the phantom center?

☐ Does it pull you ear in one direction or another?

☐ What happens to the mix buss meters?

C) Unsolo the channels and listen with all the mix elements.

☐ Does the sound lean to one side?

☐ Where is your attention focused?

☐ What happens to the mix buss meters?

D) Solo the stereo channels again. Pan them as in Exercise 6.1.

☐ Does each mix element have its own space in the stereo field?

☐ Does the mix sound like it's in stereo?

☐ What happens to the mix buss meters?

E) Unsolo the tracks and listen to the mix.

☐ Does the sound lean to one side?

☐ Does the mix still sound like it's in stereo?

☐ Does the mix have a center focus?

☐ What happens to the mix buss meters?

F) Experiment with your own pan settings. Remember to try to keep each mix element in its own space in the stereo field.

Panning The Drums

Back in the days when a drum kit was recorded with only a few microphones, the drums were recorded in mono, sometimes even on a single track. Today with most drum kits and drum loops tracked in stereo, the entire mix is built upon the idea that the drums will take up a lot of space in the sound field.

There are two ways to pan the drums. Most mixers will pan it the way they see the kit set up, with the high hat slightly on the right and floor tom on the left for a right-handed player. A few mixers choose to pan from the drummers perspective, where everything is reversed.

▶ Exercise 6.3: Panning The Overheads

Begin with the mix that you achieved for "It's About Time" in the last chapter.

A) Start by muting all the tracks except the drum channels. Pan the Overheads channel hard to the left and right.

☐ What happens to the sound?

☐ What happens to the mix buss meters?

B) Now pan the Overheads to the ten and two o'clock positions.

☐ What happens to the sound?

☐ What happens to the mix buss meters?

C) Pan the Overheads both to the center.

☐ What happens to the sound?

☐ What happens to the mix buss meters?

D) Now reverse the position of the overheads so that the left is panned hard right and the right is panned hard left.

☐ What does it sound like?

☐ Can you hear each cymbal on both sides?

☐ What happens to the mix buss meters?

▶ Exercise 6.4: Panning The Kick

A) Pan the Kick channel to the hard left.

☐ What happens to the sound?

☐ What happens to the mix buss meters?

B) Pan the Kick channel back to the center.

☐ What happens to the sound?

☐ What happens to the mix buss meters?

▶ Exercise 6.5: Panning The Snare

A) Pan the Snare subgroup or individual Snare channels to the hard right.

☐ What happens to the sound?

☐ What happens to the mix buss meters?

B) Pan the Snare channels back to the center.

☐ What happens to the sound?

☐ What happens to the mix buss meters?

C) Pan the Snare channels to about one o'clock position where it would sit if you were looking at the drummer playing.

☐ What happens to the sound?

☐ Does it sound natural?

☐ What happens to the mix buss meters?

▶ Exercise 6.6: *Panning The High-Hat*

A) Pan the Hat channel hard to the left.

☐ What happens to the sound?

☐ What happens to the mix buss meters?

B) Pan the Hat channel to the center.

☐ What happens to the sound?

☐ What happens to the mix buss meters?

C) Pan the Hat channel to about two o'clock where it would sit if you were looking at the drummer playing it.

☐ What happens to the sound?

☐ Does it sound natural?

☐ What happens to the mix buss meters?

▶ Exercise 6.7: *Panning The Toms*

Go to a place in the song with a drum fill.

A) Pan the right Tom High hard right, the Tom Low hard to the left.

☐ What does it sound like?

☐ What happens to the mix buss meters?

B) Pan the right Tom High hard left, the Tom Low hard to the right.

☐ What does it sound like?

☐ What happens to the cymbal leakage?

☐ What happens to the mix buss meters?

C) Pan the Tom High to about the two o'clock position, the Tom Low to the ten o'clock position where they would sit if you were looking at a drummer playing them.

☐ Is the sound balanced?

☐ Does it sound natural?

☐ What happens to the mix buss meters?

D) If mixing a song using a floor tom, first start with it panned to the 9 o'clock position, as you would see it on the drummer's kit.

Panning Percussion

High frequency percussion like shakers and tambourines can mask the cymbals unless panning is taken into careful consideration. Likewise, low-frequency percussion like bongos and congas can mask or be masked by other mix elements, but panning them to an open spot in the sound field can stop this from happening without any additional processing like EQ.

> **Exercise 6.8**: *Panning Percussion*

Go to the outro of the sample song where all three percussion tracks are playing.

A) First, unmute the Bongos, Tambourine and Shaker tracks. Now go to a place in the song where they all play at the same time (on the outro).

☐ What does it sound like when they're all panned to the center?

☐ Are any of the percussion tracks masked?

☐ Are any of the cymbal tracks masked?

B) Look at the panning of the Hat. If it's panned to around 2 o'clock, then pan the Shaker to the same place on the left (10 o'clock).

☐ Can you hear the Shaker better?

☐ Can you hear the Hat better?

C) Since the Tambourine is hitting on beats 2 and 4 like the Snare, pan it to the same place as the Snare but on the left (11 o'clock).

☐ Can you hear the Tambourine better?

☐ Can you hear the Snare better?

D) Pan the Bongos hard left, then try hard right.

☐ Can you hear the Bongs better?

☐ Does it draw your attention away from the other mix elements?

☐ Does it make the mix lean in the direction of the Bongos?

E) Pan the Bongos near, but slightly away from the Tambourine.

☐ Can you hear the Bongs better?

☐ Does it draw your attention away from the other mix elements?

☐ Does it make the mix lean in the direction of the Bongos?

☐ Does it sound better panned to the center?

TIP: *The idea with panning percussion is to make it balance with the drums in the sound field.*

Panning The Bass

The bass is another instrument with a lot of low frequencies that brings more power to the mix if it's panned in the center. Stereo bass is rarely, if ever, used, but if it is, it's generally panned only slightly out from center to still provide an anchor for the song.

▶ *Exercise 6.9*: *Panning The Bass*

Begin with the mix that you achieved in the last chapter.

A) Keep the drums soloed and solo the Bass. Pan it hard to the left and then to the right.

☐ Does the mix now lean to one side?

☐ Is your attention taken to where the bass is panned?

☐ What happens to the mix buss meters?

B) Return the Bass to the center.

☐ What does it sound like?

☐ What happens to the mix buss meters?

Panning Guitars

Guitars are an instrument that can benefit greatly from panning. Sometimes just moving a guitar slightly in the sound field will make the difference from it being masked or being heard.

▶ *Exercise 6.10*: *Panning Guitars*

Go to the intro of the song.

A) Unmute Electric Guitar 1. Pan it hard right.
- ☐ Can you hear it better than before?
- ☐ What happens to the mix buss meters?

B) Unmute Strat and pan it hard left.
- ☐ Can you hear it better than before?
- ☐ Does it sound unnatural?
- ☐ Does it pull you attention away from the center?
- ☐ What happens to the mix buss meters?

C) Now pan the Strat slightly more than the 11 o'clock position and Electric Guitar 1 slightly more than 1 o'clock.
- ☐ Does the mix sound more centered?
- ☐ Do you still get a sense of spaciousness?
- ☐ What happens to the mix buss meters?

D) Go to the verse and unmute both Electric Guitar 2 and the Sitar Channels. Pan Guitar 2 to approximately where Guitar 1 was. Pan the Sitar channel to where the Strat was.
- ☐ What does the mix sound like?
- ☐ Can you hear the guitars better?
- ☐ What happens to the mix buss meters?

TIP: You can pan instruments to places occupied by other mix elements if they're not playing at the same time.

E) Now go to the B section of the song and unmute both acoustic guitar channels. Pan Acoustic Guitar 1 to 10 o'clock and Acoustic Guitar 2 to 2 o'clock.
- ☐ Where can you pan them so you can hear all the mix elements best?

F) Unmute the Chorus Guitar channel and leave it in the center.
- ☐ Where can you pan it so you can hear it and the other instruments best?

G) Now go to the second verse of the song and unmute the Electric Guitar 3 channel. Pan it to the same place as Electric Guitar 1.
- ☐ Where can you pan it so you can hear it and the other instruments best?

H) Tweak or repan the guitar channels so they can be clearly heard in the stereo sound field.
- ☐ If you move the pan of the guitars around the sound field, is there a place where you can hear them best?

Panning Keyboards

Most acoustic keyboards today are tracked in stereo. Pianos, organs, upright pianos, harpsichords and most other keyboard instruments all lend themselves to a wide stereo image because it usually takes multiple mics to properly capture all their range. Just because they're recorded in stereo doesn't mean that they have to be panned that way however. Many times a stereo instrument panned in mono will sit much better in a mix.

> ### Exercise 6.11: *Panning The Keyboards*
Go to the first verse in the song.

A) Solo the Piano channel and pan it so that the low strings are panned hard left and the high strings are panned hard right.

☐ Does it sound like there's a hole in the middle?

☐ How does it sound in the mix when you unsolo it?

☐ Can you hear the piano well or is it buried in the mix?

☐ What happens to the mix buss meters?

B) Reverse the panning so that the high end is panned hard left and the low end is panned hard right.

☐ How does the mix sound?

☐ Can you hear it well or is it buried in the mix?

☐ What happens to the mix buss meters?

☐ Does it sound as natural?

C) Solo the Piano channel again and set both the high and low channels of the piano so that they're at the 10 o'clock position in the sound field. Then unsolo it.

☐ How does the mix sound?

☐ Can you hear it well or is it buried in the mix?

☐ What happens to the mix buss meters?

☐ Does it sound as natural as before?

☐ If you move the pan of the piano around the sound field, is there a place where you can hear it best?

☐ What happens to the mix buss meters?

D) Solo the Piano channel again, as well as the Rhodes and Wurlitzer channels. Pan them all hard left and hard right.

☐ Does it sound like there's a hole in the middle?

☐ Can you distinguish between them or do they sound like one keyboard?

☐ How do they sound in the mix when you unsolo them?

☐ Can you hear the keyboards well or are they buried in the mix?

☐ What happens to the mix buss meters?

E) Solo the three keyboard channels again and set the pans on the Piano to the 11 o'clock and 1 o'clock positions in the sound field.

F) Set the pans on the Wurlitzer channel to the 7 o'clock and 10 o'clock, and the Rhodes to 2 o'clock and 5 o'clock.

☐ Can you hear the 3 keyboards more distinctly now?

☐ Unsolo the channels. Can you hear them well or are they buried in the mix?

☐ What happens to the mix buss meters?

☐ Does it sound as natural as before?

☐ Can you still hear the stereo spaciousness?

☐ If you move the pan of the keyboards around the sound field, is there a place where you can hear them best?

☐ What happens to the mix buss meters?

Panning Vocals

With few exceptions, most lead vocals are always panned to the center of the mix so the focus of attention is always on them. Background vocals, however, are a different story. It's not uncommon to spread background vocals almost anywhere in the sound field in order to separate them from the vocal so they can be both heard clearly.

> ### Exercise 6:12: *Panning The Lead Vocal*

Go to the first verse in the song.

A) Pan the lead Vocal hard to the left.

☐ Do you find the effect distracting?

☐ Is it taking the attention away from the center of the mix?

☐ Can you hear the vocal well or is it buried in the mix?

☐ What happens to the mix buss meters?

B) Go to the pre-chorus of the song. Pan the lead Vocal and the Vocal Double so they're at the 10 o'clock and 2 o'clock positions in the sound field.

☐ How does the mix sound?

☐ Can you hear the vocal well or is it buried in the mix?

☐ What happens to the mix buss meters?

C) Pan the lead Vocal and the Vocal Double so that they're at the center position in the sound field.

☐ How does the mix sound?

☐ Can you hear it well or is it buried in the mix?

☐ What happens to the mix buss meters?

▶ Exercise 6.13: *Panning The Background Vocals*

Go to the B section of the song in order to hear the first set of background vocals.

A) Solo the background vocal tracks, then pan them so that they're at the center position in the sound field. Unsolo them and listen in the mix.

☐ How does the mix sound?

☐ Can you still hear the lead vocal well or is it masked by the background vocals in the mix?

☐ What happened to the mix buss meters?

B) Solo the background vocals again and pan all the 1 vocal tracks so that they're hard left, and the 3 vocal tracks hard right in the sound field. Leave the 2's in the center. Unsolo them and listen in the mix.

☐ How does the mix sound?

☐ Can you hear the background vocals well or are they buried in the mix?

☐ Can you hear the lead vocal better or is it masked?

☐ What happened to the mix buss meters?

C) Solo the background vocals again. Pan the BGv Low 1 so that it's at the 11 o'clock and and the BGv Low 3 so it's at the 1 o'clock position in the sound field. Keep the BGv Low 2 in the center.

D) Now pan the BGv Mid 1 so that it's at the 10 o'clock and and the BGv Mid 3 so it's at the 2 o'clock position in the sound field. Keep the BGv Mid 2 in the center.

E) Now pan the BGv High 1 so that it's at the 9 o'clock and and the BGv High 3 so it's at the 3 o'clock position in the sound field. Keep the BGv High 2 in the center.

☐ How do the background vocals sound?

☐ Unsolo the background vocals. How do they sound in the mix?

☐ Can you hear them well or are they buried in the mix?

☐ Can you hear the lead vocal or is it masked?

☐ If you move the pan of the background vocals around the sound field, is there a place where you can hear them best?

☐ What happened to the mix buss meters?

F) Pan the Harmony vocal channels in the same manner during the pre-chorus and chorus.

► Exercise 6.14: *Panning The Horns*

Go to the chorus of the song in order to hear the horns.

A) Solo the horn channels, then pan them so that they're at the center position in the sound field. Unsolo them and listen in the mix.

☐ Can you hear the horns well or are they buried in the mix?

☐ Do they mask any other mix elements, especially the lead vocal?

☐ What happened to the mix buss meters?

B) Solo the horn channels again and pan all the 1 vocal tracks so that they're hard left, and the 2 channels hard right in the sound field. Unsolo them and listen in the mix.

☐ Can you hear the horns well or are they buried in the mix?

☐ Can you hear the lead vocal better?

☐ What happened to the mix buss meters?

C) Solo the horns again. Pan them exactly the same way as the background vocals.

☐ Does it still fell like the horns are in stereo?

☐ Unsolo the horns vocals. How does the mix sound?

☐ Can you hear them well or are they buried in the mix?

☐ Can you hear the lead vocal?

☐ If you move the pan of the horns around the sound field, is there a place where you can hear them best?

☐ What happened to the mix buss meters?

F) Pan the horns close to, but not exactly the same as the background vocals?

☐ Can you hear both the horns and background vocals better, or is one or the other masked in the mix?

► Exercise 6.15: *Final Panning*

Go to the outro of the song where all channels are playing.

A) Look at the panning of all the channels.

☐ What happens if you alter it slightly so every channel is panned to a slightly different spot in the sound field?

☐ Are any tracks still masked?

☐ Can you hear every track distinctly? Don't worry if you can't. We'll fix that soon.

◆

Chapter 7
Dynamics Processors

Plugins and hardware devices that are designed to affect the volume envelope of a signal are known as Dynamics Processors. These can modify the envelope in a number of different ways that can change the sound a surprising amount under the right circumstances. Among the types of dynamics processors available include compressors, limiters, de-essers, and gates. Let's take a look at each.

Compression

In real life, music has a wide dynamic range that sometimes varies from an almost silent whisper to a spine tingling roar. The problem is that vinyl records, CDs, radio, television, MP3s and just about any other kind of audio distribution medium have a dynamic range that just can't accommodate these wide swings in volume that are so common in normal life. In order to fit almost any kind of audio into the limited dynamic range of those mediums, there has to be some modification of that large dynamic range into something that fits the medium, and that's a major reason why we use a compressor.

This was more essential in the days of magnetic tape and vinyl records, but more modern formats like the CD and audio files have a much wider dynamic range, so that large amount of compression was no longer a necessity. The problem is that we have become accustomed to the sound of compressed audio that we now prefer to hear it that way.

As a result, today's compressors are used not only for the essential matter of dynamics control, but also for the sound that they impart, which is an important aspect of mixing for the engineer.

Compression Basics

A compressor is nothing more than an automated level control that uses the input signal to determine the output level. Some models do this so transparently that you can't hear them working at all, while other models impart their own sound by just being inserted into the signal path. Regardless of how they sound, they all have roughly the same parameter controls and are operated the same way.

Compressor Controls

Not every compressor has the same controls, although most modern plugins do. Let's look at these typical parameter controls.

Ratio

The *Ratio* parameter controls how much the output level of the compressor will increase compared to the level being fed to the input (see Figure 7.1). For instance, if the compression ratio is set at 4:1 (four to one), that means for every 4dB of level that goes into the compressor, only 1dB will come out once the signal reaches the threshold level (the point at which the compressor begins to work). If a compression ratio is set at 8:1, then for every 8dB that goes into the unit, only 1dB will come out of the output. On some compressors, the ratio control is fixed, but on most compressors the *Ratio* parameter is variable from 1:1 (where no compression occurs) to as much as 100:1 (where it then becomes a limiter, a subject that we'll address later in this chapter). Some compressors (like the famous UREI LA-2A - see Figure 7.2 - and LA-3) have a fixed ratio that gives it a particular sound.

Figure 7.1: Typical Compressor Controls
Courtesy of PSP Audioware

Figure 7.2: A UREI/Teletronix LA-2A Compressor
Courtesy of Universal Audio

Threshold

The *Threshold* control determines the signal level where the compression begins (see Figure 7.1). Below the threshold point, no compression occurs. For instance, many compressor meters are calibrated in dB, so a setting of -5dB means that when the level reaches -5dB on the input meter, the compression begins to kick in.

Attack And Release

Most, but not all, compressors have *Attack* and *Release* parameter controls (see Figure 7.1). These controls determine how fast or slow the compressor reacts to the beginning (the attack) and body (the release) of the signal envelope. Many compressors have an *Auto* mode that automatically sets the attack and release according to the volume envelope of the signal. Although *Auto* works relatively well, it may not be precise enough to allow you to dial in the settings required by certain source material sources. Some compressors (again like the famous UREI LA-2A or the dbx 160A - see Figure 7.3) have a fixed attack and release that can't be altered, which helps give the compressor a distinctive sound. Other compressors (like the famous Fairchild 670 - see Figure 7.4) have selectable attack and release parameters.

Figure 7.3: A dbx 160A Compressor
Courtesy of dbx

Figure 7.4: The UAD Fairchild 670 Compressor Plug-in
Courtesy of Universal Audio

The *Attack* and *Release* controls are the key to proper compressor setup, but many engineers overlook these controls completely. It's possible to get good results by keeping these controls set to the mid-way or default position, but learning how to use them provides much more consistent and professional results. We'll cover this soon.

Gain, Make-Up Gain, Output

When a compressor actually compresses the signal, the output level is decreased, so there needs to be another parameter that boosts the signal back up to where it was before it was compressed. Depending upon the compressor, this parameter control may be called either *Gain, Make-Up Gain* or *Output* (see Figure 7.1). For example, if you were applying 5dB of compression, then you'd have to set the *Make-Up Gain* to +5 in order for the signal to be at the same level as when it's bypassed.

Gain Reduction Meter

The *Gain Reduction Meter* is an indicator of just how much compression is occurring at any given moment (see Figure 7.1). On most devices this is shown via a VU or peak meter that reads backwards. In other words, it's set at zero and usually travels to the left into the minus range to show compression. As an example, a meter that reads -4dB indicates that there is 4dB of compression occurring at that time (see Figure 7.5), and therefore there is 4dB less level at the output if no make-up gain is applied.

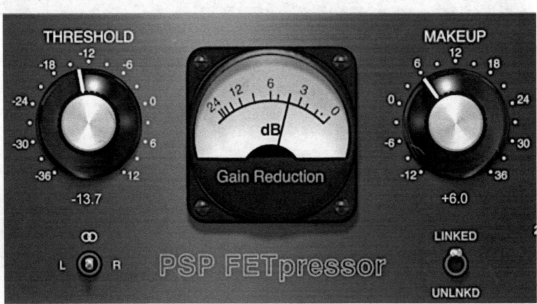

Figure 7.5: An Example Of 4dB Of Compression
Courtesy of PSP Audioware

The Side Chain

Many compressors also have an additional input and output called a side chain, which is used for connecting other external signals to it (see Figure 7.6). This allows the compressor to receive a signal from another track or signal source that will make it trigger into gain reduction.

A good example is the sidechain input of a compressor used on a bass track that is connected to the kick drum channel. When the kick drum hits, it will trigger the compressor, which will reduce the gain of the bass channel making the kick drum stand out more.

You can connect a delay, reverb or anything you want to side chain for unusual, program level-dependent effects. A side chain isn't needed for normal compressor operations, so many manufacturers and software developers choose not to include it on their hardware and plugins.

Figure 7.6: FabFilter Pro C-2 sidechain controls (bottom)
Courtesy of FabFilter

Bypass

Most compressors, especially most of the plugin versions, have a *Bypass* control that allows you to hear the signal without any gain reduction taking place. This is useful to help you hear how much the compressor is controlling or changing the sound, or to make it easy to set the *Output* control so the compressed signal is the same level as the uncompressed signal.

Compressor Operation

Controlling dynamics means keeping the level of the sound even by lifting the level of the soft passages and lowering the level of the loud ones so that there's less of a difference between them. This is typically accomplished by compressing anywhere from 2 to 6dB or so at anywhere from a 2:1 to 8:1 ratio, although some situations may require more radical settings.

Setting The Compression Ratio

The ratio parameter of a compressor is actually easy to set if you follow a quick rule of thumb. If you want an instrument to have more punch and attack, set the ratio low, to 1.5:1, 2:1 or 3:1. This allows more of the signal transients to get through. If you're looking for more control of the signal, set the ratio higher, like 4:1 to 10:1 or even more.

TIP: The higher the ratio parameter is set, the more likely you'll hear the compressor work, which may not be desirable.

Setting The Attack And Release

The setting of the attack and release is important to making the compressor work as intended, so here are a few steps for set up. One of the easiest ways to do that is to use the snare drum as your template, assuming that you're mixing a song with a more or less constant tempo, then use the same approximate *Attack* and *Release* settings for the other instruments. *The idea is to make the compressor breathe with the pulse of the song.*

1. Start with the attack time set as slow as possible (usually all the way to the right), and release time set as fast as possible (usually all the way to the left) on the compressor.

2. Turn the attack faster until the instrument (in this case, the snare) begins to sound dull (this happens because you're compressing the attack portion at the beginning of the sound envelope). Stop increasing the attack time at this point and even back it off a little.

3. Adjust the release time so that after the snare hits, the volume goes back to 90 to 100 percent normal by the next snare beat.

4. Add the rest of the mix back in and listen. Make slight adjustments to the attack and release times as needed.

TIP: Since there are different types of compressors and they all work differently, the same parameter settings probably won't sound the same from model to model or developer to developer.

How Much Compression Do I Need?

How much compression you use is a matter of taste. That being said, the more compression you use, the more likely that you'll hear it working. Generally speaking, compression of 6dB or less is used more for controlling dynamics than for imparting any sonic quality, but it's also common to see as much as 15 or even 20dB used for electric guitars, room mics, drums, and even vocals, depending upon the situation. In the final analysis, the amount of compression depends on the song, the arrangement, the player, the room, the instrument or vocalist, or the sound you're looking for.

Compression As An Effect

Compression is interesting because of how much it can change the sound of a track under the right circumstances. Sometimes it can make a track seem closer to the listener, or seem more aggressive and exciting. The *Attack* and *Release* controls can modify the volume envelope of a sound to have more or less attack or release, which can make it sound punchy or fatter, or make a note have a longer decay.

Sometimes massive amounts of compression (like 15 or 20dB) can impart a sound into the track that you can't get any other way, and sometimes even a dB or two can change the sound of a track just enough to get you where you want to go.

Limiting

While a compressor increases the low level and decreases the loud ones to even out the dynamic range, a limiter keeps the level from ever going much louder once it hits the threshold point when the limiter turns on. It's very much like a truck with a speed governor on it that keeps the truck at 60 mph regardless of how much more you press down on the gas pedal. With a limiter, once you hit the predetermined signal level, it never gets much louder no matter how much more input level it receives.

A compressor and a limiter are somewhat identical except for the settings. Any time the compression ratio is set to 10:1 or more, it's considered a limiter. Limiting is usually used in sound reinforcement for speaker protection (there are often limiters on powered studio monitors as well), and not used as much as compressors in mixing with a few exceptions.

Most modern digital "brickwall" limiters (either hardware or plugin) have a function known as "look ahead" which allows the detector circuit to look at the signal a millisecond or two before it hits the limiter. This means that the limiter acts extremely fast and just about eliminates any overshoot of the predetermined level, which essentially stops digital overs from occurring. See Figure 7.7 for an example of a digital limiter with a look-ahead function.

TIP: A look-ahead type limiter can be identified by a Ceiling parameter, which is the absolute highest level that the signal will go. It's often used in on the master mix buss during mixing and during mastering.

Figure 7.7: The PSP Twin-L Brickwall Limiter
Courtesy of PSP Audioware

Many engineers who feel that the bass guitar is the anchor for the song want the bass to have as little dynamic range as possible. This can be achieved by limiting the bass by 3 to 6dB (depending on the song) with a ratio of 10:1, 20:1 or even higher or by using a brickwall limiter. Many engineers prefer to use this on the kick and snare as well.

Compressing The Various Instruments

Most instruments can benefit from at least some compression, either for the purpose of smoothing out a performance, raising or lowering a note or drum hit that might be getting lost or is too loud, or as an effect where it adds color to the sound. Let's look at how it works on the most commonly used mix elements.

Compressing The Drums

There are a number of reasons to compress the drums. Sometimes a drummer doesn't hit every beat on the kick and snare with the same intensity, which makes the pulse of the song feel erratic. Sometimes the toms fills on the same track have different volumes with each hit. And in terms of effects, compression does work wonders to push the kick and snare forward in the track to make them much more punchy. Let's do some experimenting.

▶ *Exercise 7.1: Learning The Compressor*

Using the setup and tracks from the previous chapter on "Hard Plastic," insert a compressor on any insert of the Claps channel. To begin, set the *Threshold* control as high as it will go so that no compression occurs, set the *Ratio* set at 2:1, and the *Attack* and *Release* controls and the *Output* control set to their default positions. The object is to change the color and punch of the track.

A) Solo the Claps channel, then slowly adjust the *Threshold* until the *Gain Reduction Meter* reads 3dB.

☐ Can you hear the compression?

☐ Did the sound change?

☐ Did the level change?

☐ Can you hear a difference if you bypass the compressor?

B) Adjust the *Threshold* until the *Gain Reduction Meter* reads 10dB.

☐ Can you hear the compression?

☐ How much did the level change?

☐ Can you hear a difference if you bypass the compressor?

C) Return the *Threshold* control to where there's only 3dB of gain reduction. Now adjust the *Ratio* control from 2:1 to 8:1.

☐ What does the gain reduction meter read now?

☐ What does the master mix buss meter read now?

☐ Can you hear the compression?

☐ Can you hear a difference if you bypass the compressor?

D) Now increase the *Ratio* control from 8:1 to 20:1.

☐ What does the gain reduction meter read now?

☐ What does the master mix buss meter read now?

☐ Can you hear the compression?

☐ Can you hear the difference if you bypass the compressor?

E) Return the *Ratio* control to 2:1 and adjust the *Threshold* control until there's about 3dB of gain reduction occurring. Now adjust the *Attack* time to as fast as it will go (all the way to the right on most compressors).

☐ What does the gain reduction meter read now?

☐ Has the level changed?

☐ Can you hear the compression?

☐ Can you hear the difference if you bypass the compressor?

F) Increase the *Attack* time to as fast as it will go (all the way to the left on most compressors).

☐ What does the gain reduction meter read now?

☐ Can you hear the compression?

☐ Can you hear the difference if you bypass the compressor?

G) Now adjust the *Attack* time to as slow as it will go (all the way to the right on most compressors)..

☐ What does the gain reduction meter read now?

☐ Can you hear the compression?

☐ Can you hear the difference if you bypass the compressor?

G) Decrease the *Attack* time until the sound of the Claps just begins to dull. Decrease the *Threshold* to maintain 3dB of compression if needed.

☐ What does the gain reduction meter read now?

☐ Can you hear the compression?

☐ Can you hear the difference if you bypass the compressor?

H) Increase the *Release* time to as slow as it will go.

☐ What does the gain reduction meter read now?

☐ Can you hear the compression?

☐ What did you notice about how the gain reduction meter reacts?

☐ Can you hear the difference if you bypass the compressor?

I) Decrease the *Release* time to as fast as it will go.

☐ What does the gain reduction meter read now?

☐ What did you notice about how the gain reduction meter reacts?

☐ Can you hear the compression?

☐ Can you hear the difference if you bypass the compressor?

J) Increase the *Release* time until the gain reduction meter seems to pulse with the track (about 300 to 400ms).

☐ What does the gain reduction meter read now?

☐ Can you hear the compression?

☐ Can you hear the difference if you bypass the compressor?

K) Select the *Bypass* to hear the volume of the Kick without the compressor.

☐ Now deselect the *Bypass* and slowly raise the *Output* or *Gain* control until the compressed signal level is equal to the uncompressed signal in level. Continue to use the *Bypass* to check.

If the amount of gain reduction is 4dB, for example, then start by adding 4dB of gain on the output control.

☐ Make sure that the level touched -10dB on the master meters.

L) Unsolo the Claps channel and listen to it with the rest of the mix.

☐ What do you notice about the Claps?

☐ Can you hear the compression?

☐ What does the master mix buss meter read now?

M) Try a different compressor.

☐ Can you hear a difference?

☐ Do you like the sound better?

▶ Exercise 7.2: Compressing A Synth Loop

We can really see and hear a compressor at work by compressing the Active Pad channel on "Hard Plastic." This track has wide changes of level, so the aim of the compressor is to even those level changes out a bit.

A) Solo the Active Pad channel. Set the *Ratio* at about 4:1 and the *Threshold* to where there's about 3dB of compression.

☐ Did the sound of the loop change?

☐ Are the notes more even?

☐ Can you hear the compression?

☐ Can you hear any difference in the sound if you bypass the compressor?

B) Add some makeup gain so that the level is about the same regardless of whether the compressor is inserted or bypassed.

☐ Can you hear the compression?

☐ What does the master mix buss meter read now?

☐ Can you hear the difference if you bypass the compressor?

B) With the channel still soloed, decrease the *Threshold* until the *Gain Reduction Meter* reads 10dB. Be sure to add makeup gain so that the level is the same when bypassed or inserted.

☐ Can you hear the compression?

☐ What does the master mix buss meter read now?

☐ Can you hear the difference if you bypass the compressor?

C) Return the *Threshold* control to where there's only 3dB of gain reduction. Now increase the *Ratio* control to 8:1.

☐ What does the master mix buss meter read now? Is it more even when the loud and soft phrases play?

☐ Can you hear the compression?

☐ Can you hear the difference if you bypass the compressor?

☐ Does the loop sound more even or does it sound choked?

D) Set the *Attack* and *Release* controls as described previously with the aim of keeping the level as stead as possible, but also experiment with extreme settings.

☐ What happened when you set the *Attack* set as long as possible?

☐ What happened when you set the *Release* set as long as possible?

☐ What happened when you set the *Attack* set as short as possible?

☐ What happened when you set the *Release* set as short as possible?

E) Bypass the compressor and listen to the level.

☐ Does it need any make-up gain? If so, add the same amount as the level is reduced on the gain reduction meter.

☐ Is the level about the same in with the compressor in or out of the signal chain?

F) Try a different compressor.

☐ Can you hear a difference?

☐ Do you like the sound better?

G) With the attack and release set, unsolo the Active Pad channel.

☐ What does it sound like in the mix?

☐ Is it more or less present?

☐ Can you hear it better?

☐ Does the channel level need adjusting?

H) Add compression to the Bright Digi and Pad channels as needed.

▶ *Exercise 7.3: Compressing The Kick*

Using the setup and tracks from your previous mix of "It's About Time," insert a compressor on any insert of the kick drum channel. To begin, set the *Threshold* control as high as it will go so that no compression occurs, set the *Ratio* set at 2:1, and the *Attack* and *Release* controls and the *Output* control set to mid-way.

A) Solo the Kick, then slowly adjust the *Threshold* until the *Gain Reduction Meter* reads 3dB. Set the *Ratio* control to 2:1.

☐ Can you hear the compression?

☐ Did the sound change?

☐ Did the level change?

☐ Can you hear a difference if you bypass the compressor?

B) Adjust the *Threshold* until the *Gain Reduction Meter* reads 10dB.

☐ Can you hear the compression?

☐ How much did the level change?

☐ Can you hear a difference if you bypass the compressor?

C) Return the *Threshold* control to where there's only 3dB of gain reduction. Now adjust the *Ratio* control from 2:1 to 8:1.

☐ What does the gain reduction meter read now?

☐ What does the master mix buss meter read now?

☐ Can you hear the compression?

☐ Can you hear a difference if you bypass the compressor?

D) Now increase the *Ratio* control from 8:1 to 20:1.

☐ What does the gain reduction meter read now?

☐ What does the master mix buss meter read now?

☐ Can you hear the compression?

☐ Can you hear the difference if you bypass the compressor?

E) Return the *Ratio* control to 2:1 and adjust the *Threshold* control until there's about 3dB of gain reduction occurring. Now adjust the *Attack* time to as fast as it will go (all the way to the right on most compressors).

☐ What does the gain reduction meter read now?

☐ Has the level changed?

☐ Can you hear the compression?

☐ Can you hear the difference if you bypass the compressor?

F) Increase the *Attack* time to as slow as it will go (all the way to the left on most compressors).

☐ What does the gain reduction meter read now?

☐ Can you hear the compression?

☐ Can you hear the difference if you bypass the compressor?

G) Now adjust the *Attack* time to as fast as it will go again, the decrease the *Attack* time until the sound of the Kick just begins to dull.
- ☐ What does the gain reduction meter read now?
- ☐ Can you hear the compression?
- ☐ Can you hear the difference if you bypass the compressor?

H) Increase the *Release* time to as slow as it will go.
- ☐ What does the gain reduction meter read now?
- ☐ Can you hear the compression?
- ☐ What did you notice about how the gain reduction meter reacts?
- ☐ Can you hear the difference if you bypass the compressor?

I) Decrease the *Release* time to as fast as it will go.
- ☐ What does the gain reduction meter read now?
- ☐ What did you notice about how the gain reduction meter reacts?
- ☐ Can you hear the compression?
- ☐ Can you hear the difference if you bypass the compressor?

J) Increase the *Release* time until the gain reduction meter stays seems to pulse with the track.
- ☐ What does the gain reduction meter read now?
- ☐ Can you hear the compression?
- ☐ Can you hear the difference if you bypass the compressor?

K) Select the *Bypass* to hear the volume of the Kick without the compressor.
- ☐ Now deselect the *Bypass* and slowly raise the *Output* control until the compressed signal level is equal to the uncompressed signal in level. Continue to use the *Bypass* to check. If the amount of gain reduction is 4dB, for example, then start by adding 4dB of gain on the output control.
- ☐ Make sure that the level touched -5dB on the master meters.

L) Unsolo the Kick track and listen to it with the rest of the mix.
- ☐ What do you notice about the Kick?
- ☐ Can you hear the compression?
- ☐ What does the master mix buss meter read now?

M) Try a different compressor.
- ☐ Can you hear a difference?
- ☐ Do you like the sound better?

▶ Exercise 7.4: *Compressing Other Drums*

Follow the same procedure as in Exercise 7.3 to hear the extreme effects of the parameter settings.

> **NOTE: Be careful with the Snare of "It's About Time" since the compression won't react as it normally might because of all the ghost notes. The Claps track on "Hard Plastic" is a better example to practice on.**

▶ Exercise 7.5: *Compressing The Room Mics*

A) Solo the room mics and follow the same procedure as A through F in 7.1 to hear the extreme effects of the parameter settings.

B) Set the *Attack* time as slow as it will go, then increase it until the sound of the room just begins to dull.

☐ What does the gain reduction meter read now?

☐ Can you hear the compression?

☐ Can you hear the difference if you bypass the compressor?

C) Increase the *Release* time until the gain reduction meter seems to breathe with the tempo of the track. *NOTE: Make sure the gain reduction returns to zero and doesn't stay continually lit.*

☐ What does the gain reduction meter read now?

☐ Can you hear the compression?

☐ Can you hear the difference if you bypass the compressor?

D) Select the *Bypass* to hear the volume of the room without the compressor.

☐ Now deselect the *Bypass* and slowly raise the *Output* control until the compressed signal is equal to the uncompressed signal in level.

☐ Continue to use the *Bypass* to check.

E) Solo the other drum tracks and listen to the room tracks along with them.

☐ What do you notice about the room?

☐ Can you hear the compression?

☐ Does does the drum mix change when you bypass the compressor?

F) Many mixers prefer the room sound to be extremely compressed. Solo the Drum Room Channel and increase the compression by increasing the *Threshold* or *Ratio* controls until there's about 10dB or more of compression, then tuck the room tracks in just under the other drum tracks.

☐ What do you notice about the drum sound?

☐ If you solo all the drum tracks, what do you notice?

☐ What happens to the sound of the drum mix when you bypass the compressor?

G) With all drum tracks soloed, slowly increase the fader level of the Drum Room channel until it hits -10.

☐ What do you notice about the sound of the drums?

☐ Does does the drum mix change when you bypass the compressor?

H) Try a different compressor.

☐ Can you hear a difference?

☐ Do you like the sound better?

▶ Exercise 7.6: *Compressing The Entire Kit*

A) Insert a compressor into the drum subgroup signal path so it will effect the entire drum kit. Set the *Ratio* on the compressor to 2:1 and adjust the *Threshold* so there's about 3dB of gain reduction.

☐ What do you notice about the drum sound?

☐ Did you notice a change in the kick and snare?

B) Set the *Attack* time as slow as it will go, then increase it until the sound of the kit just begins to dull and the punch of the sound increases.

☐ What does the gain reduction meter read now?

☐ Can you hear the compression?

☐ Can you hear the difference if you bypass the compressor?

☐ What does it do to the cymbals?

C) Now increase the *Release* time to as slow as it will go, then decrease it so it seems to breathe with the tempo of the track as in previous exercises.

☐ What does the gain reduction meter read now?

☐ What does the master mix buss meter read now?

☐ Can you hear the difference on the kick and snare?

☐ Can you hear the difference if you bypass the compressor?

☐ What do the cymbals sound like?

D) Select the *Bypass* to hear the volume of the room without the compressor.

☐ Now deselect the *Bypass* and slowly raise the *Output* control until the compressed signal is equal to the uncompressed signal in level.

☐ Continue to use the *Bypass* to check both the level and if the compression is actually adding to the sound.

E) Increase the *Ratio* control to 10:1.

☐ What does the gain reduction meter read now?

☐ What does the master mix buss meter read now?

☐ Can you hear the compression?

☐ Can you hear the difference if you bypass the compressor?

☐ What do the cymbals sound like?

E) Return the Ration control to 4:1 and increase the compression 10 dB by decreasing the *Threshold*.

☐ What does the gain reduction meter read now?

☐ What does the master mix buss meter read now?

☐ Can you hear the compression?

☐ Can you hear the difference if you bypass the compressor?

☐ What do the cymbals sound like?

F) Bypass the subgroup drum compressor.

☐ Can you hear a difference?

☐ Does it sound as punchy?

G) Try a different compressor.

☐ Can you hear a difference?

☐ Do you like the sound better?

Parallel Compression

There's a great trick that really punches up the drum sound without adding more compression to the individual tracks. It's something I call the "New York Compression Trick" because when I was starting out, every mixer that I knew that worked in New York City used it on their mixes. Now everyone uses it so it's not that exclusive to New York any more, so we'll just call it by its more academic name - parallel compression (although the real New York City Compression Trick is a little more complicated than that).

Essentially the trick centers around an additional drum subgroup that has a compressor with some rather extreme settings. Once the the subgroup is set up and compressor is kicking, the subgroup is gently raised until it's just barely heard against the original drum mix. If you want the drums punchier, just add more subgroup level.

Be warned - the sound that you get out of the drums when using parallel compression is addicting, and you'll want to use it on every mix (which is perfectly okay if it works for you).

▶ Exercise 7.7: Parallel Compression On The Drums

A) Assign the drums to separate subgroup. If you're already bussed to one, use a second separate one. *Make sure you bypass any compression you have on the drum subgroup.*

B) Insert a stereo compressor to the subgroup and set it so there's about 10dB of compression and the *Attack* and *Release* so it breathes with the track.

C) Raise the fader level of the subgroup with the compressor -10 on the fader. it's tucked just under the present rhythm section mix to where you can just hear it.
☐ Does it sound punchier?

☐ What do the master mix buss meters read?

D) Adjust the fader level of the subgroup with the compressor until it's tucked just under the present rhythm section mix to where you can just hear it.
☐ Does it sound punchier?

☐ What do the master mix buss meters read?

E) For an even greater effect, EQ the subgroup with +6dB at 10kHz and +6dB at 100Hz.
☐ What does it sound like now?

F) Try a different compressor.
☐ Can you hear a difference?

☐ Do you like the sound better?

Compressing The Bass

Most electric bass guitars inherently have notes that are louder or softer than others depending upon where they're played on the neck of the instrument. This is especially noticeable on a bass played with a pick instead of with the fingers. Some notes just roar while others might get lost, which is why at least some compression is usually necessary on the instrument. On the other hand, there are some engineers that build their mixes around the bass and want the level to be virtually the same throughout the song, so they'll set the compressor accordingly.

The *Ratio* control is important to dialing in the right amount of compression on the bass. Watch the channel meter and if there are a lot of wild peaks, a higher ratio and higher threshold (which provides less compression,) is required. If you just want to round out the sound, use a lower compression ratio and a lower threshold for more compression.

▶ Exercise 7.8: Compressing The Bass

Solo the bass. If there are direct and amp tracks and they're subgrouped, then solo the subgroup and insert a compressor in the subgroup signal path.

A) Set the *Ratio* at 4:1 and adjust the *Threshold* so that there's about 3dB of compression.

☐ Did the sound of the bass change?

☐ Are the notes more even?

☐ Can you hear the compression?

☐ What happens to the master mix buss meters?

B) Decrease the *Threshold* until the *Gain Reduction Meter* reads 10dB. Make sure to adjust the makeup gain so the level is the same as when it's bypassed.

☐ Can you hear the compression?

☐ Is the bass smoother sounding? Does it have any peaks that aren't controlled?

☐ What does the master mix buss meter read now?

☐ Can you hear the difference if you bypass the compressor?

C) Return the *Threshold* control to where there's only 3dB of gain reduction. Now increase the *Ratio* control from 4:1 to 12:1.

☐ What does the gain reduction meter read now?

☐ What does the master mix buss meter read now?

☐ Can you hear the compression?

☐ Can you hear the difference if you bypass the compressor?

☐ Does the bass sound more even or does it sound choked?

D) Return the *Ratio* control to 4:1 and increase the *Threshold* control until there's about 3dB of gain reduction occurring. Now decrease the *Attack* time to as fast as it will go.

☐ What does the gain reduction meter read now?

☐ What does the master mix buss meter read now?

☐ Can you hear the compression?

☐ Can you hear the difference if you bypass the compressor?

☐ Does the bass sound better or worse?

E) Increase the *Attack* time to as slow as it will go.

☐ What does the gain reduction meter read now?

☐ What does the master mix buss meter read now?

☐ Can you hear the compression?

☐ Can you hear the difference if you bypass the compressor?

☐ Does the bass sound better or worse?

F) Now decrease the *Attack* time until the sound of the bass just begins to dull.

☐ What does the gain reduction meter read now?

☐ What does the master mix buss meter read now?

☐ Can you hear the compression?

☐ Can you hear the difference if you bypass the compressor?

☐ Does the bass sound better or worse?

G) Set the *Release* time to as slow as it will go.

☐ What does the gain reduction meter read now?

☐ What does the master mix buss meter read now?

☐ Can you hear the compression?

☐ Can you hear the difference if you bypass the compressor?

H) Decrease the *Release* time to as fast as it will go.

☐ What does the gain reduction meter read now?

☐ What does the master mix buss meter read now?

☐ Can you hear the compression?

☐ Can you hear the difference if you bypass the compressor?

☐ Does the bass sound better or worse?

I) Increase the *Release* time until the bass breathes with the track (makes the notes feel longer), or set it to the approximate *Release* setting of the snare compressor.

☐ What does the gain reduction meter read now?

☐ What does the master mix buss meter read now?

☐ Can you hear the compression?

☐ Can you hear the difference if you bypass the compressor?

J) Select the *Bypass* to hear the volume of the bass without the compressor. Now deselect the *Bypass* and slowly raise the *Output* control until the compressed signal is equal to the uncompressed signal in level. Continue to use the *Bypass* to check.

K) Unsolo the Bass channel or bass subgroup and listen to it with the rest of the mix.

☐ What do you notice about the bass?

☐ Can you hear the compression?

☐ What does the master mix buss meter read now?

L) Many mixers prefer that the bass have no dynamics in order to keep the mix solid and punchy. To accomplish this, increase the *Ratio* control to 12:1 or more and increase the *Threshold* until there is between 3 and 6dB of compression.

☐ Can you hear the compression?

☐ Do you like this sound better?

☐ Does the bass fit in the track better?

☐ Does it still have any peaks?

M) Try a different compressor.

☐ Can you hear a difference?

☐ Do you like the sound better?

The settings for a miked bass amp might be different, depending upon how distorted it is.

Compressing Guitars

Clean electric guitars and acoustic guitars can greatly benefit from compression, but distorted guitars are already naturally compressed. That being said, a little extra compression can make a lead guitar stand out in a mix.

Acoustic and clean electric guitars generally have the a lot dynamic range and usually require more compression. Direct clean guitars require the most, sometimes 10dB or more. With a guitar that's amplified, usually the more distorted it becomes, the less compression it requires, although most electric guitars can benefit from at least a few dB.

As with the bass, the *Ratio* control is important to dialing in the right amount of compression. Watch the meter and if there are a lot of wild peaks, a higher ratio and higher threshold is required. If you just want to round out the sound, use a lower compression ratio and a lower threshold, which will give you more compression.

▶ Exercise 7.9: Compressing Electric Guitars

A) Go to the intro of the song and solo the Strat channel. Set the *Ratio* at about 4:1 and the *Threshold* to where there's about 3dB of compression.

☐ Did the sound of the guitar change?

☐ Are the notes more even?

☐ Can you hear the compression?

B) Decrease the *Threshold* until the *Gain Reduction Meter* reads 10dB.

☐ Can you hear the compression?

☐ What does the master mix buss meter read now?

☐ Can you hear the difference if you bypass the compressor?

C) Return the *Threshold* control to where there's only 3dB of gain reduction. Now increase the *Ratio* control from 4:1 to 8:1.

☐ What does the gain reduction meter read now?

☐ What does the master mix buss meter read now?

☐ Can you hear the compression?

☐ Can you hear the difference if you bypass the compressor?

☐ Does the bass sound more even or does it sound choked?

D) Set the *Attack* and *Release* controls as described previously to breathe with the track, but also experiment with extreme settings.

☐ How did the sound change with the *Attack* set as long as possible?

☐ How did the sound change with the *Release* set as long as possible?

☐ How did the sound change with the *Attack* set as short as possible?

☐ How did the sound change with the *Release* set as short as possible?

E) Bypass the compressor and listen to the level.

☐ Does it need any make-up gain? If so, add the same amount as the level is reduced on the gain reduction meter.

☐ Is the level about the same in with the compressor in or out of the signal chain?

F) With the attack and release set, unsolo the Channel.

☐ What does it sound like in the track?

☐ Is it more or less present?

☐ Can you hear it better?

☐ Does the channel level need adjusting?

G) Try a different compressor.

☐ Can you hear a difference?

☐ Do you like the sound better?

H) Try a compressor on all the other guitar channels in the song and see if and how it makes a difference.

> **Exercise 7.10**: *Compressing The Acoustic Guitar*

A) Go to the B section of the song and solo the Acoustic Set the Ratio at about 4:1 and the Threshold to where there's about 3dB of compression.

☐ Did the sound of the guitar change?

☐ Are the notes more even?

☐ Can you hear the compression?

B) Decrease the *Threshold* until the *Gain Reduction Meter* reads 10dB.

☐ Can you hear the compression?

☐ What does the master mix buss meter read now?

☐ Can you hear the difference if you bypass the compressor?

C) Return the *Threshold* control to where there's only 3dB of gain reduction. Now increase the *Ratio* control from 4:1 to 8:1.

☐ What does the gain reduction meter read now?

☐ What does the master mix buss meter read now?

☐ Can you hear the compression?

☐ Can you hear the difference if you bypass the compressor?

☐ Does the acoustic guitar sound more even or does it sound choked?

B) Set the *Attack* and *Release* controls as described previously to breathe with the track, but also experiment with extreme settings.

☐ How did the sound change with the *Attack* set as long as possible?

☐ How did the sound change with the *Release* set as long as possible?

☐ How did the sound change with the *Attack* set as short as possible?

☐ How did the sound change with the *Release* set as short as possible?

C) Bypass the compressor and listen to the level.

☐ Does it need any make-up gain? If so, add the same amount as the level is reduced on the gain reduction meter.

☐ Is the level about the same in with the compressor in or out of the signal chain?

D) With the attack and release set, unsolo the Channel.

☐ What does it sound like in the track?

☐ Is it more or less present?

☐ Can you hear it better?

☐ Does the channel level need adjusting?

E) Try a different compressor.

☐ Can you hear a difference?

☐ Do you like the sound better?

F) Now solo both Acoustic 1 and Acoustic 2.

☐ Do they sound different?

☐ Try adding a compressor to Acoustic 2 (or just copy it from Acoustic 1) and see if the same settings work

G) Unsolo both and listen to the mix.

Compressing Keyboards

Like guitars, compression on keyboards depends on how wild the dynamic swings are. An acoustic piano is inherently much more dynamic than a synthesizer or organ, so it must be treated differently as a result.

Sampled acoustic or electric pianos don't have nearly the dynamic range of a real acoustic instrument, but they still can have some major peaks depending upon the way they're played. Beware that the more compression used, the less realistic an acoustic piano (or virtual piano) sounds.

Sometimes organ and string sounds, which aren't very dynamic, can benefit from a touch of compression to make sure all the notes are heard evenly, which also pulls it in front of the mix a bit.

▶ *Exercise 7.11*: *Compressing Keyboards*

A) Solo the Wurlitzer channel. Set the *Ratio* at about 4:1 and the *Threshold* to where there's about 3dB of compression.

☐ Did the sound of the keyboard change?

☐ Are the notes more even?

☐ Can you hear the compression?

B) Decrease the *Threshold* until the *Gain Reduction Meter* reads 10dB.

☐ Can you hear the compression?

☐ What does the master mix buss meter read now?

☐ Can you hear the difference if you bypass the compressor?

C) Return the *Threshold* control to where there's only 3dB of gain reduction. Now increase the *Ratio* control to 8:1.

- ☐ What does the gain reduction meter read now?
- ☐ What does the master mix buss meter read now?
- ☐ Can you hear the compression?
- ☐ Can you hear the difference if you bypass the compressor?
- ☐ Does the Keyboard sound more even or does it sound choked?

B) Set the *Attack* and *Release* controls as described previously to breathe with the track, but also experiment with extreme settings.

- ☐ What happened when you set the *Attack* set as long as possible?
- ☐ What happened when you set the *Release* set as long as possible?
- ☐ What happened when you set h the *Attack* set as short as possible?
- ☐ What happened when you set the *Release* set as short as possible?

C) Bypass the compressor and listen to the level.

- ☐ Does it need any make-up gain? If so, add the same amount as the level is reduced on the gain reduction meter.
- ☐ Is the level about the same in with the compressor in or out of the signal chain?

D) Try a different compressor.

- ☐ Can you hear a difference?
- ☐ Do you like the sound better?

E) With the attack and release set, unsolo the Wurlitzer channel.

- ☐ What does it sound like in the track?
- ☐ Is it more or less present?
- ☐ Can you hear it better?
- ☐ Does the channel level need adjusting?

F) Add compression to the other keyboard channels as needed.

Compressing Vocals

If there's one mix element that greatly benefits from compression it's the human voice. Most singers aren't able to sing every word or line at the same level, so some words get buried as a result. Compression evens out the level differences so you can better hear every word.

The amount of compression can vary wildly on a vocal if it has a lot of dynamic range, with a whisper to a scream within the same song, so it's not uncommon to use as much as 10dB or more on some vocals.

▶ Exercise 7.12: Compressing The Lead Vocal

A) Solo the lead Vocal channel. Set the *Ratio* at about 4:1 and the *Threshold* to where there's about 3dB of compression.

☐ Did the sound of the vocal change?

☐ Are the words and phrases more even in level?

☐ Can you hear the compression?

B) Set the *Attack* and *Release* controls as described previously to breathe with the track, but also experiment with extreme settings.

☐ What happened when you set the *Attack* set as long as possible?

☐ What happened when you set like with the *Release* set as long as possible?

☐ What happened when you set the *Attack* set as short as possible?

☐ What happened when you set the *Release* set as short as possible?

C) Bypass the compressor and listen to the level.

☐ Does it need any make-up gain? If so, add the same amount as the level is reduced on the gain reduction meter.

☐ Is the level about the same in with the compressor in or out of the signal chain?

D) With the attack and release set, unsolo the lead vocal.

☐ What does it sound like in the mix?

☐ Is it more or less present?

☐ Can you hear it better?

☐ Does the channel level need adjusting?

E) Try a different compressor.

☐ Can you hear a difference?

☐ Do you like the sound better?

F) Increase the compression by lowering the *Threshold*.

☐ Does the vocal come out front more?

☐ Does it sound too compressed or choked?

Compressing Other Mix Elements

The principles of compressing other mix elements are identical to any of the above instruments, so let's review.

- Acoustic instruments are usually more dynamic than electric or virtual instruments, and therefore need to be controlled more.

- The more wild the peaks, the higher the compression ratio should be set. The fewer the peaks, the lower the ratio.

- The more compression you use, the more likely that you'll hear it and color the sound.

- If you set the attack and release times of the compressor so it breathes with the track, the less likely you'll hear it working in the mix.

Compressing Loops And Samples

Care should be taken when compressing loops, beats, or samples as these are usually heavily compressed already. One of the byproducts of additional compression is that it's possible to change the groove, which won't be desirable in most cases. Sometimes just a few dB of limiting can handle the peaks and allow it to sit better in the mix though.

▶ *Exercise 7.13*: *Compressing Loops, Samples And Beats*

A) Go to the song "Hard Plastic" and solo the Active Pad.

- **Set the Ratio at about 2:1**

- **Set the Threshold to where there's about 3dB of compression.**

- **Set the *Attack* and *Release* controls to breathe with the track as in the previous exercises.**

- **Set the *Gain* control to make up for the gain reduction as in previous exercises.**

☐ Are the notes and phrases more even in level?

☐ Unsolo it and see how it fits in the track?

☐ Can you hear the compression?

☐ Does the groove change?

B) Increase the compression to 10dB.

☐ What does it sound like?

☐ Unsolo it and see how it fits in the track?

☐ Can you hear the compression?

☐ Does the groove change?

C) Set the Ratio at about 12:1 and the *Threshold* to where there's about 3dB of compression.

☐ What does it sound like?

☐ Unsolo it and see how it fits in the track?

☐ Can you hear the compression?

☐ Does the groove change?

D) Increase the compression to 10dB.

☐ What does it sound like?

☐ Unsolo it and see how it fits in the track?

☐ Can you hear the compression?

☐ Does the groove change?

E) Set the *Ratio* and the amount of compression that seems to work best within the mix and balance it accordingly.

F) Try a different compressor.

☐ Can you hear a difference?

☐ Do you like the sound better?

De-essers

Sometimes a vocal has short bursts of high-frequency energy where the "S's" are over-emphasized, which is known as "sibilance." It comes from a combination of mic technique by the vocalist, the type of mic used, and heavy compression on the vocal track. Sibilance is nasty sounding and generally undesirable, so a special type of compressor called a a de-esser is used (see Figure 7.8). A de-esser can be tuned to compress only a selected band of frequencies between 3kHz and 8kHz to eliminate sibilance.

Figure 7.8: The de-esser in the Scheps Omni Channel plugin

Most de-essers only have only two controls; *Threshold* and *Frequency*. Some have a *Listen* button that allow you to solo only the frequency that's being compressed, which can be helpful in finding the offending frequency.

If you have a de-esser or de-esser plugin, try the following exercise.

> ▶ **Exercise 7.14**: *Using The De-esser*

> **A) Solo the lead vocal and insert the de-esser on the channel *after the compressor*.**

> **B) Lower the *Threshold* control until the sibilance is decreased, but so you can still hear the "S,s". If you can't hear them, then you've lowered the *Threshold* too far.**

> **C) Span the available frequencies with the *Frequency* control until you find the one that's most offensive.**

> **D) Un-solo the vocal and listen in context of the mix.**

> ☐ Is the sound of the vocal natural?

> ☐ Is there any remaining sibilance?

> ☐ Can you distinguish the "S's"?

Gates

A gate (sometimes called "Noise Gate") is sort of a reverse-compressor. That is, it works backwards from a normal compressor in that the sound level is at its loudest until it reaches the threshold, where it's then attenuated or muted completely.

A gate can be used to cover up noises, buzzes, coughs or other low level noises that were recorded on a track. For example, a gate can be used on electric guitar tracks to effectively eliminate amplifier noise when the guitar player isn't playing. On drums, gates can be used to turn off the leakage from the tom mics, since that tends to muddy up the other drum tracks.

Like the de-esser, a gate can sometimes consist of just a few controls, principally the *Threshold, Range* and sometimes *Hold* or *Release* controls (see Figure 7.9). *Range* sets the amount of attenuation after the threshold is reached and the gate turns on. The *Hold* control keeps the gate open a defined amount of time, and the *Release* control works just like on a compressor, controlling how much of the tail of the sound we'll hear before the gate closes.

TIP: Sometimes when gating drums, the Range control is set so it attenuates the signal only about 10 or 20dB. This lets some of the natural ambience remain and prevents the drums from sounding choked.

Figure 7.9: A SSL Duende gate
Courtesy of Solid State Logic

▶ *Exercise 7.15: Using A Gate On The Toms*

Go to a place in the song where there's a Tom Low fill. Solo the Tom Low and insert the gate on the channel.

A) Raise the *Threshold* control until you can hear the tom hit, but no sound in between hits. Un-solo the track.

☐ Does it sound natural?

☐ Does it sound cut off?

B) Solo the track again and adjust the *Range* control so the Tom Low channel is attenuated by 10dB to 20dB between hits.

☐ Does it sound more natural?

☐ Does it sound cut off?

C) If the gate stutters before off and on, try fine tuning the settings of the *Release* and *Hold* controls (if the gate has one). If it still stutters, add a compressor *before* the gate to keep the signal steady.

☐ Does the tom sound cut off? If so increase the *Release* or *Hold* time.

☐ Does it sound natural when played with the other drums?

☐ Does it sound natural when played with the full mix? If not go back and fine tune the *Release* and *Hold* controls.

D) If satisfied with the sound, copy to the other tom channels. You might have to fine tune each for the best results.

Chapter 8
Using The EQ

No part of mixing draws more questions that adding EQ. Where do I do it? How much is enough? What frequencies do I use? Is it okay if I don't use any? These are all legitimate inquiries that many beginning mixers have that will be answered as this chapter goes along.

Equalization Basics

Before you begin twisting equalizer knobs, it helps to understand the situations where EQing might be necessary, since there are more than one. They are:

- To make a mix element sound clearer and more defined.
- To make a mix element sound bigger and larger than life.
- To make it smaller so it fits in the track better.
- To make all the elements of a mix fit together better.

It's all too common for someone new to mixing to solo a track, grab the EQ, and endlessly search for what seems to be the right sound. The only problem is that you don't know what the right sound really is unless you listen to everything else in the mix together.

If you keep these four goals in mind, you'll avoid a lot of indiscriminate knob twirling and endless searching for the right setting.

TIP: Remember, there's no rule that says you have to use the EQ at all if the track sounds good and works fits into the mix already.

EQ Parameters

There are a number of parameters that you'll find on most equalizers (see Figure 8.1).

Figure 8.1: The controls of a typical equalizer

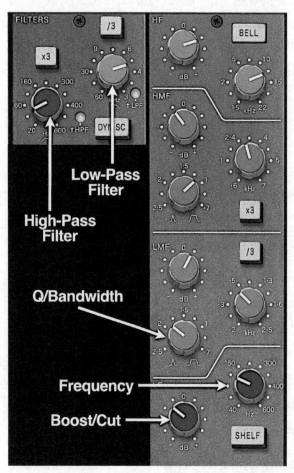

Frequency - The *Frequency* control selects the center frequency around which the equalizer operates. It comes in many forms. It can be a fixed frequency like the tone controls on a car radio or guitar amplifier, selectable frequencies where a button or detent selects the frequency, or a variable or swept frequency control where you can continuously select the frequency that's appropriate.

Boost/Cut - This is the control that adds or subtracts the volume of the particular frequency, or band of frequencies, chosen.

Shelving/Bell Curve - Many high and low frequency equalizers (not mid-range) have the ability to switch between a shelving or bell curve. Shelving attenuates or boosts all frequencies above or below a specified cutoff point. These are the common Bass and Treble controls that you find on consumer audio devices. A Bell curve boosts a frequency and its surrounding frequencies at a set point.

Q (Bandwidth) - This control selects the band of frequencies that the equalizer will boost or cut. A Q set on 10 will only affect a narrow number of frequencies around the frequency selected, so it's very precise and therefore excellent for cutting an offensive frequency spike. A Q of 2 affects several octaves above and below the center frequency (if 1,000Hz is selected, it may effect everything from 500Hz through 2,000Hz or more). The *Q* or *Bandwidth* control is not found on all equalizers.

High-Pass Filter - The high-pass filter (sometimes abbreviated with HPF) allows high frequencies to pass and attenuates low frequencies. Sometimes the HPF (also sometimes more appropriately called "Low-Cut" because the low frequencies are attenuated) has a fixed frequency like at 40 or 60Hz; sometimes there are several frequencies that are selectable; and sometimes the frequency selection is continuously variable.

Low-Pass Filter - The low-pass filter (sometimes abbreviated with LPF) allows low frequencies to pass and attenuates the high frequencies. Sometimes the LPF (also sometimes more appropriately called "High-Cut" because the high frequencies are attenuated) has a fixed frequency like 10k or 12kHz; sometimes there are several frequencies that are selectable; and sometimes the frequency selection is continuously variable.

In/Out - *In* inserts the equalizer into the signal path while *Out* bypasses it.

A Description Of The Audio Bands

The audio frequency spectrum is made up of six basic yet distinct frequency bands. Each one has an enormous impact of the final sound, so it's important to know the characteristics of each before we begin to use the equalizer.

Figure 8.2: Audio Band Description

Frequency Band	Description	Consequences
Sub-Bass 16Hz to 60Hz	Sounds that are often felt more than heard. They give the music a sense of power	Too much emphasis in this range makes the music sound muddy. Attenuating this range (especially below 40Hz) can clean a mix up considerably.
Bass 60Hz to 250Hz	Contains the fundamental notes of the rhythm section	EQing this range can make the musical balance sound either fat or thin. Too much boost in this range can make the music sound boomy.
Low Mids 250Hz to 2kHz	Contains the low harmonics of most musical instruments	Can introduce a telephone-like quality to the music if boosted too much. Boosting the 500 to 1000Hz octave makes the instruments sound horn-like. Boosting the 1k to 2kHz octave makes them sound tinny. Excess output in this range can cause listening fatigue.
High Mids 2k to 4kHz	Controls the speech recognition sounds of m, b and v	Too much boost in this range, especially at 3kHz, can introduce a lisping quality to a voice. Too much boost in this range can cause listening fatigue. Dipping the 3kHz range on instrument backgrounds and slightly peaking 3kHz on vocals can make the vocals audible without having to decrease the instrumental level in mixes where the voice would otherwise seem buried.
Presence 4k to 6kHz	Responsible for the clarity and definition of voices and instruments	Boosting this range can make the music seem closer to the listener. Reducing the 5kHz content of a mix makes the sound more distant and transparent.
Brilliance 6k to 16kHz	Controls the brilliance and clarity of sounds	Too much emphasis in this range can produce sibilance to the vocals.

▶ *Exercise 8.1*: *Getting Familiar With The Equalizer*

A) Begin by using the mix of "It's About Time" that you have started and soloing only the Snare Top channel.

☐ Can you hear the snare distinctly?

☐ Is it crisp, full, thin or dull sounding?

B) Add 6dB at 1kHz.

☐ How did the sound change? If you can't hear a change turn the monitor level up a little louder.

☐ Does it sound closer to you?

☐ Does it sound better or worse when you bypass the EQ?

C) Cut 6dB at 1kHz.

☐ How did the sound change?

☐ Does it sound closer to you?

☐ Does it sound better or worse when you bypass the EQ?

☐ Did it loose its presence?

D) Return the EQ to 0dB, the add 6dB at 5kHz.

☐ How did the sound change?

☐ Does it sound closer to you?

☐ Does it sound better or worse when you bypass the EQ?

☐ Does it sound more or less clear and distinct with the EQ in?

E) Now cut 6dB at 5kHz.

☐ How did the sound change?

☐ Does it sound closer to you?

☐ Does it sound better or worse when you bypass the EQ?

☐ Does it sound more or less clear and distinct with the EQ in?

F) Return the EQ to 0dB, the add 6dB at 8kHz using a shelving curve.

☐ How did the sound change?

☐ Does it sound closer to you?

☐ Does it sound better or worse when you bypass the EQ?

☐ Does it sound more or less clear and distinct with the EQ in?

G) Now cut 6dB at 8kHz using a shelving curve.

☐ How did the sound change?

☐ Does it sound closer to you?

☐ Does it sound better or worse when you bypass the EQ?

☐ Does it sound more or less clear and distinct with the EQ in?

H) Return the EQ to 0dB, the add 6dB at 8kHz using a bell curve.

☐ How did the sound change?

☐ Does it sound closer to you?

☐ Does it sound better or worse when you bypass the EQ?

☐ Does it sound more or less clear and distinct with the EQ in?

I) Now cut 6dB at 8kHz using a bell curve.

☐ How did the sound change?

☐ Does it sound closer to you?

☐ Does it sound better or worse when you bypass the EQ?

☐ Does it sound more or less clear and distinct with the EQ in?

J) Return the EQ to 0dB, then add 6dB at 125Hz using a shelving curve.

☐ Does it sound fuller?

☐ Does it sound muddy?

K) Now cut 6dB at 125Hz using a shelving curve.

☐ How did the sound change?

☐ Did it loose its fullness?

L) Return the EQ to 0dB, then add 6dB at 125Hz using a bell curve.

☐ Does it sound fuller?

☐ Does it sound muddy?

M) Now cut 6dB at 125Hz using a bbell curve.

☐ How did the sound change?

☐ Did it loose its fullness?

N) Experiment with using different amounts of EQ at those frequencies.

O) Experiment on other mix elements.

The Magic High-Pass Filter

One of the most useful and overlooked equalization parameters available for a mixer is the high-pass filter (HPF). The high-pass filter can be another parameter as part of an equalizer, or it can be a stand-alone plugin or device. The HPF does just what it says; it allows high frequencies to pass and cuts off low frequencies.

The low frequencies of many instruments sometimes just clash with each other and in the end, don't add much to the sound anyway. That's why if you attenuate the low frequencies below 100 or even 150Hz on most mix elements other than the kick and bass, the mix begins to clean up almost magically.

For instance, by rolling off the low frequencies of a vocal track, you can eliminate the rumble of trucks and machinery that you can't physically hear because they're so low, yet they were recorded and can therefore muddy up the mix. Rolling off the low end of an electric guitar keeps it out of the way of the rhythm section and helps it to fit better the mix. Even rolling off the bass or drums anywhere below 40 to 60Hz can sometimes make the mix both louder and punchier without any sense of losing the low end.

There's only so much room in the frequency band of a mix, so the more mix elements the smaller each on must be in order to fit. The high-pass filter can also be used as a way to slim down a track to make it fit better without resorting to other types of EQ.

▶ *Exercise 8.2*: Using The High-Pass Filter

Let's start with the example song "Hard Plastic." There are a number of tracks that can benefit from a high-pass filter since they don't have or need a lot of low-frequency energy. Let's start with the Hat channel.

A) Solo the Hat channel and insert an EQ plugin that has a high-pass filter built in, or a dedicated filter plugin.

B) Set the filter Q or Slope to 12dB/oct and the frequency to 200Hz.

☐ Does the sound change when you bypass the plugin?

☐ Does it sound a little thinner with the filter in?

C) Increase the frequency until you can hear the sound change, then solo the Kick and Claps channels.

☐ Does the Hat fit in better with the Kick and Claps now?

☐ Does it sound clearer when you listen to the entire mix?

D) Set the high pass filter on the Hat channel to where you think it sounds best, unsolo, the solo the Pad channel. Insert insert an EQ plugin that has a high-pass filter built in, or a dedicated filter plugin on the channel.

E) Set the filter Q or Slope to 12dB/oct and the *Frequency* to 200Hz.

☐ Does the sound change when you bypass the plugin?

☐ Can you hear that the low-frequency noise is eliminated when the filter is inserted?

☐ Does the Pad sound thinner with the filter in?

F) Increase the *Frequency* until you just hear the sound change.

☐ Can you still hear any low-frequency noise?

☐ Does the Pad sound clearer when you listen to the entire mix?

▶ *Exercise 8.3*: Using The High-Pass Filter On Low-Frequency Mix Elements

The approach to adding a high-pass filter is different for low-frequency mix element like kick, bass or floor tom. Filtering out too many low-frequencies can make the element lose girth and power, but something it can also keep two low-frequency elements from clashing. For this exercise, we'll be listening to the entire mix.

A) Insert an EQ plugin that has a high-pass filter built in, or a dedicated filter plugin on the Kick channel. Set the filter Q or *Slope* to 12dB/oct.

B) Set the *Frequency* to 100Hz.

☐ Does it sound a little thinner with the filter in?

☐ Does the sound change when you bypass the plugin?

C) Set the *Frequency* to 80Hz.

☐ Does it sound a little thinner with the filter in?

☐ Does the sound change when you bypass the plugin?

D) Set the *Frequency* to 60Hz.

☐ Does it sound a fuller now?

☐ Does it fit better in the mix?

E) Set the *Frequency* to 40Hz.

☐ Does it sound a fuller now?

☐ Does it fit better in the mix?

F) Set the *Frequency* to 20Hz.

☐ Does it sound a fuller now?

☐ Does it sound different or bigger when bypassed?

G) Insert an EQ plugin that has a high-pass filter built in, or a dedicated filter plugin on the Bass channel. Set the filter Q or *Slope* to 12dB/oct.

H) Set the *Frequency* to 100, 80, 60, 40 and 20Hz.

☐ Does the bass fit the mix better?

☐ Does the sound change when you bypass the plugin?

I) Choose the high-pass filter frequency for the Kick and Bass that you think sounds best. If in doubt choose 20Hz for the Kick and 40Hz for the bass, or vice versa.

Using The Equalizer

There are many different methods for EQing an instrument or vocal mix element, but here are a couple of tried and true techniques that will never steer you wrong.

Subtractive Equalization

While it's natural to believe that by adding some EQ here and there that you'll make the mix element sound better, that's not necessarily the case.

There's a very effective EQ technique called "subtractive equalization" that works by attenuating frequencies instead of boosting them. Many superstar mixers love this method because it makes the sound of the track more natural than if you boosted any of the frequencies.

This is because every time you boost an EQ, there's a slight amount of something called phase shift that's added to the signal as a byproduct of the way an electronic or digital equalizer works. By using subtractive equalization, you lessen the affects of this artifact. As a result, the track is better able to blend with the others.

Here's an example of how to use subtractive equalization:

1. Set the *Boost/Cut* control of a midrange equalizer band to a moderate level of *boost* (8 or 10dB should work.)

2. Set the *Q* control (if there is one) to a higher number like 8 or 10.

3. Sweep through the frequencies until you find the frequency that really jumps out above the others. That's the frequency to *cut*.

4. Adjust the amount of cut to taste. Be aware that too much cut makes the sound thinner.

There are three frequency ranges that are particularly effective when using subtractive equalization; from 100 to 200Hz, from 400 to 600Hz, and between 2k and 4kHz. The reason why 100 to 200Hz is chosen is because most directional microphones provide a natural boost in that frequency range because of the proximity effect brought about by miking an instrument or voice up close.

Likewise, many mics that are known as good vocal mics have a presence boost between 2k and 4kHz. Cutting those frequencies a few dB (more or less as needed) can make the track sound much more natural than if you were to try to boost other frequencies instead.

The 400 to 600Hz region is where the sound becomes "boxy." With a kick drum it could provide a sound closer to a beach ball being hit. Cutting a few dB here can sometimes clean up the sound considerably.

These three problem areas usually crop up when you're recording everything with the same microphone, since there's a buildup in the those frequency areas as more and more instruments are recorded. By cutting a few dB from these frequency ranges you'll find that the instruments sit better in the mix without every having to add as much from the other EQ bands.

Juggling Frequencies

One of the biggest problems during mixing is when two instruments clash because their predominant frequencies are in the same bandwidth area. A great example of this is when two similar sounding guitars are in the mix (like if they're both Strats played through Marshalls), but lots of other combinations, like between a guitar and a lead vocal, or a snare drum and a guitar, happen as well. The way to avoid this to use a method called "juggling frequencies."

TIP: Veteran engineers know that soloing a track when equalizing it without listening to any of the other tracks at the same time inevitably may cause a frequency clash between the EQed track and another track in the mix. The way to avoid this is to listen to other instruments while you're EQing, and when you find two instruments that have frequencies that clash, solo only those.

Here's how frequency juggling is done.

- Make sure that the equalizers of the two offending channels are not boosting at the same frequency. If they are, move one to a slightly higher or lower frequency.

- If a mix element's equalizer is cut at a certain frequency, boost the frequency of the other mix element at that same frequency or vice-versa. For example, if the kick is cut at 500Hz, boost the bass at 500Hz.

Also, you'll probably have to do a lot of back-and-forth EQing where you start with one instrument, then tweak another that's clashing, then return to the original one, and back again over and over until each one can be heard distinctly.

TIP: Remember that after frequency juggling, an instrument might sound terrible when soloed by itself. That's okay, because the idea is for it to work well in the track with the rest of the instruments, not sound good alone.

► *Exercise 8.4*: EQing For Definition

This exercise will use a combination of subtractive EQ, frequency juggling and high-pass filter. **PLEASE NOTE:** *The EQ settings supplied may be a bit more radical than what might be normally used so it's easy to hear the difference.*

Begin by using the mix of "It's About Time" that you have started, starting from the last chorus of the song. Solo the entire drum kit either individually or via the group or subgroup.

A) First we'll work on the Kick. Insert an EQ into the Kick channel and set it as follows (remember that you're listening to the entire drum kit, not just the Kick channel):

- **High-Pass filter: 40Hz at 12dB/oct**
- **Low or Low-Mid EQ set to bell curve: -4dB at 400Hz with a Q of 1**
- **Mid-Range EQ: +5dB at 2.8kHz with a Q of 1.6**

☐ Can you hear the Kick distinctly?

☐ Is it crisp and full sounding, or thin and dull?

☐ Can you hear the difference when you bypass the EQ?

☐ When you bypass the individual bands, can you hear the difference?

☐ Do you see how the high-pass filter, subtractive EQ, and EQing for definition worked?

B) Let's do the same with the snare. Insert an EQ into the Snare subgroup channel and set it as follows (remember that you're listening to the entire drum kit):

- **High-Pass filter: 80Hz at 12dB/oct**
- **Low or Low-Mid EQ set to bell curve: +3dB at 150Hz with a Q of 1**
- **Mid-Range EQ: -4dB at 800Hz with a Q of 1**
- **High or High-Mid EQ: +4dB at 3.5kHz with a Q of 1**

☐ Can you hear the Snare distinctly?

☐ Is it crisp and full sounding, or thin and dull?

☐ Can you hear the difference when you bypass the EQ?

☐ When you bypass the individual bands, can you hear the difference?

☐ Do you see how the high-pass filter, subtractive EQ, and EQing for definition worked?

C) Solo only the Hat channel. Insert an EQ on the channel, set the High-Pass Filter to 12dB/oct and begin to raise the frequency until sound of the Hat changes.

☐ What happed to the leakage?

☐ Did the sound of the Hat change?

D) Solo the rest of the drum kit. On the Hat channel, add the high-frequency EQ using the bell curve until the you can hear the hat distinctly in the mix.

☐ Does it come forward in the mix?

☐ Does it sound tinny or too bright?

E) Are there any other channels where the EQ is boosted at the same frequency?

• **If so, solo only that channel and the Hat to hear if they're conflicting**

• **If there is a conflict, move the frequency on the equalizer above or below the other until the conflict is removed.**

☐ Can you hear each channel distinctly now?

☐ Can you hear them both in the mix?

F) Solo the Tambourine channel.

• **Do the same the with the high-pass filter than you did with the Hat**

• **Now solo the Hat and listen to both together**

• **Set the EQ until you can hear both clearly using the previous techniques. Be careful that you don't boost a the same frequency as the Hat.**

☐ Can you hear both the Hat and Tambourine clearly when you solo just the drums?

☐ Can you hear both the Hat and Tambourine clearly unsolo all channels and listen to the mix?

G) EQ the rest of the drum kit and percussion using the technique you used above.

☐ Can you hear all the drum and percussion mix elements clearly in the mix?

☐ Did any of the mix elements loose the heft or presence?

EQing Various Instruments

While many engineers have certain frequencies that they either like or don't like, it's much easier to get everything to fit together in a mix if you realize that every mix is different. The song is different, the players are different, the instruments might be different, the arrangement is different for sure, and a whole host of other things that truly make each one unique. That's what we both love and hate about music. It's not a cookie cutter operation at all, although you'll find that each type of mix element may have certain frequency ranges that work a little better than others.

Let's look at equalizing some commonly used instruments.

Equalizing The Drums

The drums present an interesting dilemma - does the song demand that the drum kit work as a whole, or should the snare or kick stand out? Once again, it depends upon the song and the genre of music, but we can take a look at both approaches.

The kick and snare are extremely important in modern music because the kick is the heartbeat and the snare drives the song. By simply getting the sound and balance of these two drums right, it's possible to change a song from dull to exciting.

There are certain frequencies on different drums that you should be aware of.

Kick - Bottom at 80 to 100Hz, beach ball effect or hollowness at 400Hz, definition at 3k to 5kHz

Snare - Fatness at 120 to 240Hz, point at 900Hz, crispness at 5kHz, snap at 10kHz

Hat - Clang at 200Hz, sparkle at 8k to 10kHz

Rack Toms - Fullness at 240 to 500Hz, attack at 5k to 7kHz

Floor Tom - Fullness at 80Hz, attack at 5kHz

Cymbals - Clang at 200Hz, sparkle at 8k to 10kHz

These frequencies are not cut and dried for each drum kit, drum loop or drum sample, since the size of the drum or cymbal and the material it's made of contributes greatly to the tone. *Remember to sweep around each of the above frequencies to find the correct area to EQ that particular drum, cymbal or sample.*

Beware that boosting from 40 to 60Hz may make the kick sound big on your speakers, but it might not be heard when played back on smaller speakers.

TIP: The frequency area for a typical 22 inch kick drum (which is the most commonly used) is around 80 to 100Hz.

Equalizing The Bass Mix Element

The bass mix element provides the power to the mix and the fundamental to the chord changes, but it's the relationship between it and drum elements that really makes a mix sound big and fat. That's why some mixers will spend hours just trying to fine tune this balance, because if the relationship isn't correct, then the song will just never sound powerful.

At its simplest, EQing the bass element (either bass guitar or synthesizer) at a higher frequency (like 120Hz) and the kick at a lower one (like 80Hz) or vice-verse can work, however it's usually a matter of using the frequency juggling method of EQing to find the frequencies that work the best.

TIP: The most common EQ frequencies for bass mix element are girth at 40-60Hz, anywhere from 100 to 250Hz for bottom, attack at 700Hz, and finger snap at 2.5kHz.

▶ *Exercise 8.5: EQing The Bass*

A) Solo the bass and drums and raise the monitor level so it's a little louder than what you'd usually listen at.

☐ Can you hear the bass well?

☐ Does it mask the kick?

☐ Can you hear each note distinctly?

B) Solo the kick and the bass.

☐ Can you hear the bass well?

☐ Does it mask the kick?

☐ Can you hear each note distinctly?

☐ Does the bass mask the kick?

C) Wherever you cut a frequency on the kick, boost that frequency on the bass. In other words, if you cut the kick at 400Hz, boost the bass by 2dB increments in this spot.

☐ Can you hear the bass and kick distinctly?

☐ Do they reinforce one another?

D) If the kick is boosted at 80Hz, boost the bass at 100 or 120Hz.

☐ Can you hear the bass and kick distinctly?

☐ Do they reinforce one another?

E) What happens if you switch these frequencies? In other words, if you boost the kick at 120Hz and the bass at 80Hz.

☐ Can you hear the bass and kick distinctly?

☐ Do they reinforce one another?

F) Does it sound fuller if you add 2dB at 60Hz on the bass channel?

☐ Does it sound muddy?

☐ Does it mask the kick?

G) Insert the high pass filter on the bass channel and set it to 60Hz.

☐ Is the bass more distinct?

☐ What happens if you move the frequency to 80Hz?

☐ What happens if you move it to 40Hz?

H) Does the bass fit better with the kick if you boost the bass at around 700Hz?

☐ Does it fit better if you add 2dB at around 2.5kHz?

☐ Is it more distinct sounding?

I) Add the rest of the drum kit.

☐ Can you hear the bass distinctly?

☐ Does it mask any other drum?

☐ Does it reinforce the kick drum?

TIP: *If the bass still isn't heard distinctly, make sure that it's not boosted at the same frequency as any of the drums, or just raise the level a bit.*

Equalizing The Vocal

The vocal is almost always the focal point of the song, so not only is it important that it's heard well in the mix, but it has to sound good too. That being said, what works on a male vocal won't necessarily work on a female.

EQing can make a vocalist sound up close and in your face, or back in the mix, so it depends on the song and the arrangement before you choose the frequencies that best work for the song.

A small boost at 125 to 250Hz can make the voice sound a bit more "chesty." Boosting between 2k to 5kHz accentuates the consonants and adds presence, which makes the vocal seem closer to the listener. The frequencies that cause sibilance are anywhere from 4k to 7kHz (they also add presence), and what's known as "air" is between 10k to 15kHz.

Feature Frequencies
- 125 - 250Hz..........chest (male)

- 4k - 7kHz..............presence, sibilance

- 10k - 15kHz..........air

TIP: For background vocals, cutting a few dB at 2k to 5kHz can separate them from the lead vocal, if that's what the song requires.

▶ Exercise 8.6: EQing The Lead Vocal

A) Listen to the vocal against the entire mix.

☐ Is the sound of the vocal too thick or too thin?

B) If the vocal seems too big or thick, cut a few dB at around 125Hz for a male, and around 250Hz for a female.

☐ Does it fit better in the mix now?

☐ Does it sound thin?

C) If the vocal needs to be thicker or bigger sounding, add a few dB at those frequencies on the vocal channel.

☐ Does it fit better in the mix now?

☐ Does it sound too thick?

D) If the "P's", "F's", and "S's" are not distinct enough, add a dB or two between 4k and 7kHz on the vocal channel.

☐ Does it fit better in the mix now?

☐ Does it sound strident or sibilant?

E) If the "S's" are too bright:

☐ Try cutting a dB or two between 4k and 7kHz.

TIP: Cut too much from the vocal at 4k to 7kHz and the vocal will lose intelligibility. Use a de-esser instead to attenuate sibilance.

F) To brighten the vocal, add 2dB at 10kHz on the vocal channel.

☐ What does it sound like?

☐ If you can't hear it do anything, then add a little more.

G) Add 2dB at 12kHz instead.

☐ What does it sound like?

☐ Does it fit better in the mix now?

☐ Does it sound tinny?

H) Add 2dB at 15kHz instead.

☐ What does it sound like?

☐ Can you hear it at all?

I) Insert a HPF at 60Hz on the vocal channel.

☐ Is the vocal cleaner?

☐ Is the mix cleaner?

☐ Does the vocal sound different?

☐ Does it lack bottom?

J) Insert a HPF at 100Hz on the vocal channel.

☐ Is the vocal cleaner?

☐ Is the mixer cleaner?

☐ Does the vocal sound different?

☐ Does it lack bottom?

▶ *Exercise 8.7:* *EQing The Lead Vocal - Part 2*

A) Listen to the vocal against the entire mix.

☐ Is there another instrument clashing against it frequency-wise?

☐ If so, solo the vocal and the instrument that's clashing with it.

B) Look at the EQ of both vocal and conflicting tracks that are soloed.

☐ Is there a boost at the same spot?

☐ If so, decrease the boost or even cut at that frequency. Can you hear both tracks more distinctly?

☐ Can you hear the vocal better?

☐ Does it fit better in the mix now?

☐ Does it sound thin?

C) Alternately, look at the EQ of both tracks again.

☐ Is there a boost at the same spot?

☐ If so, adjust the frequency of the boost on the track conflicting with the vocal either up or down a little.

☐ Can you hear both tracks more distinctly?

☐ Can you hear the vocal better?

☐ Is the EQ cut on the vocal track the same as where it's boosted on the offending track? If so, move the frequency of the offending track up or down a little.

☐ Can you hear both tracks more distinctly?

☐ Can you hear the vocal better?

▶ Exercise 8.8: EQing The Lead Vocal - Part 3

A) If there are no frequency boosts or cuts that are conflicting, but a conflict still exists, listen closely to both channels.

☐ Is the conflict in the high or low frequencies?

TIP: If the conflict is in the lower frequencies, insert a HPF in the signal path of the offending channel and sweep the frequencies from 60 to 200Hz to find the right frequency.

☐ Is the conflict still there?

☐ If not, is the sound of the instrument still acceptable?

B) Alternately, insert a HPF in the signal path of the vocal channel and sweep the frequencies from 60 to 200Hz.

☐ Is the conflict still there?

☐ If not, is the sound of the vocal still acceptable?

☐ What happens if the HPF is inserted on both channels?

C) If the conflict is in the higher frequencies, set a mid-frequency EQ to 6dB of cut on the vocal channel, and sweep the frequencies between 1k and 5kHz.

☐ Is there a spot where the conflict goes away?

☐ If so, decrease the cut just to the point where you can hear the frequency conflict, then back it off a bit.

☐ Can you hear both channels distinctly?

☐ Do they both sound acceptable?

D) Un-solo both channels and listen with the rest of the track.

☐ Can you hear both tracks distinctly?

☐ Are there any other conflicts?

☐ If so, solo the tracks that are conflicting and repeat steps A through D.

▶ *Exercise 8.9*: *EQing The Lead Vocal - Part 4*

A) Listen to the vocal against the entire mix.

☐ Does the vocal seem forward enough in the mix?

B) Add 2dB to the vocal track at between 2k and 5kHz.

☐ Does it seem to come forward in the mix?

C) What happens if you add more than 2dB?

☐ Does it fit better in the mix now?

☐ Does it sound thin?

▶ *Exercise 8.10*: *EQing the Background Vocals*

If the background vocals are conflicting with the lead vocal, solo all vocal and background vocal channels.

A) On the background vocal tracks, set a mid-frequency EQ to 6dB of cut, and sweep the frequencies between 2k and 5kHz.

☐ Is there a spot where the conflict goes away?

☐ If so, decrease the cut just to the point where you can hear the frequency conflict, then back it off a bit.

☐ Can you hear both channels distinctly?

☐ Do they both sound acceptable?

B) If the conflict is in the lower frequencies, insert a HPF in the signal path of the offending channel and sweep the frequencies from 60 to 200Hz.

☐ Is the conflict still there?

☐ If not, is the sound of the instrument still acceptable?

Equalizing The Electric Guitar

Electric guitars, whether they be clean or distorted, are very dependent upon how they sit in the track with other mix elements in order to be heard in the mix. In some cases, like when a distorted guitar is playing power chords, it may be better for the guitar to blend in with the rest of the instruments rather than be heard distinctly, while at other times you want to be sure to hear every note.

Feature Frequencies

• 240 - 500Hz..........fullness

• 150Hz...................filter below

• 8kHz......................filter above

▶ *Exercise 8.11*: EQing The Electric Guitar

A) Listen to the guitar against the entire mix.

☐ Is there another instrument clashing against it frequency-wise?

☐ If so, solo the guitar and the instrument that's clashing with it.

B) Look at the EQ of both tracks.

☐ Is there a boost at the same spot if each track?

☐ If so, decrease the boost or even cut at that frequency.

☐ Can you hear both tracks more distinctly?

☐ Can you hear the vocal better?

C) Alternately, look at the EQ of both tracks again.

☐ Is there a boost at the same spot?

☐ If so, adjust the frequency of the track conflicting with the vocal either up or down a little.

☐ Can you hear both tracks more distinctly?

☐ Can you hear the guitar better?

☐ Is the EQ cut on the vocal track where it's boosted on the offending track?

☐ If so, move the frequency up or down a little.

☐ Can you hear both tracks more distinctly?

☐ Can you hear the guitar better?

▶ *Exercise 8.12*: EQing The Electric Guitar - Part 2

A) If there are no boosts or cuts that are conflicting but the frequency conflict still exists, listen closely to both channels.

☐ Is the conflict in the high or low frequencies?

B) If the conflict is in the lower frequencies, insert a HPF in the signal path of the offending channel and sweep the frequencies from 60 to 200Hz.

☐ Is the conflict still there?

☐ If not, is the sound of the instrument still acceptable?

C) Alternately, insert a HPF in the signal path of the guitar channel and sweep the frequencies from 60 to 200Hz.

☐ Is the conflict still there?

☐ If not, is the sound of the guitar still acceptable?

☐ What happens if the HPF is inserted on both channels?

E) If the conflict remains in the lower frequencies, set a mid-frequency EQ to 6dB of cut, and sweep the frequencies between 100Hz and 1kHz.

☐ Is there a spot where the conflict goes away?

☐ If so, decrease the cut just to the point where you can hear the frequency conflict, then back it off a bit.

☐ Can you hear both channels distinctly?

☐ Do they both sound acceptable?

F) If the conflict is in the higher frequencies, set a mid-frequency EQ to 6dB of cut, and sweep the frequencies between 1k and 5kHz.

☐ Is there a spot where the conflict goes away?

☐ If so, decrease the cut just to the point where you can hear the frequency conflict, then back it off a bit.

☐ Can you hear both channels distinctly?

☐ Do they both sound acceptable?

G) Unsolo both channels and listen with the rest of the track.

☐ Can you hear both tracks distinctly?

☐ Is there any other conflicts?

☐ If so, solo the tracks that are conflicting and repeat steps A through D.

Equalizing The Acoustic Guitar

The acoustic guitar has an entirely different sound from an electric guitar and therefore has to be approached differently. Plus, each acoustic guitar has its own sound depending upon the body size and tone wood that it's made from.

For instance, the bigger the guitar body (like a dreadnought or jumbo size), the more bottom end it will have, which might sound great live but could muddy up a recording. On the other hand, a small body or cutaway acoustic guitar will have less low end but that may cause it to sit better in a track as a result. The same goes for the wood that it's made from. A guitar made from rosewood will have much more body, while one made of mahogany may sound thinner, yet record better.

Like other instruments, there are certain frequencies to look at when EQing an acoustic. 80Hz will provide fullness, while you'll hear more of the body at 240Hz. The presence of the instrument can be found between 2k to 5kHz, 5k to 8kHz will make it cut through a mix, while 10kHz will accentuate any finger noises.

Feature Frequencies

- 80Hz....................fullness

- 240Hz ....body

- 2kHz - 5kHz ....presence

- 5kHz - 8kHz...........cut through the mix

- 10kHz ....finger noise

▶ Exercise 8.13: EQing The Acoustic Guitar

A) Listen to the acoustic guitar against the entire mix.

☐ Is there another instrument clashing against it frequency-wise?

☐ If so, solo the guitar and the instrument that's clashing with it.

B) Look at the EQ of both tracks that are clashing.

☐ Is there a boost at the same spot?

☐ If so, decrease the boost or even cut at that frequency.

☐ Can you hear both tracks more distinctly?

☐ Can you hear the guitar better in the mix?

C) Alternately, look at the EQ of both tracks again.

☐ Is there a boost at the same spot?

☐ If so, adjust the frequency of the track conflicting with the vocal either up or down a little.

☐ Can you hear both tracks more distinctly?

☐ Can you hear the acoustic guitar better in the mix?

☐ Can you hear both tracks more distinctly?

▶ Exercise 8.14: EQing The Acoustic Guitar - Part 2

A) If there are no boosts or cuts that are conflicting yet a frequency conflict still exists, listen closely to both channels.

☐ Is the conflict in the high or low frequencies?

B) If the conflict is in the lower frequencies, insert a HPF in the signal path of the offending channel and sweep the frequencies from 60 to 200Hz.

☐ Is the conflict still there?

☐ If not, is the sound of the instrument still acceptable?

C) Alternately, insert a HPF in the signal path of the acoustic guitar channel and sweep the frequencies from 60 to 100Hz.

☐ Is the conflict still there?

☐ If not, is the sound of the guitar still acceptable?

☐ What happens if the HPF is inserted on both channels?

D) If the conflict remains in the lower frequencies, set a mid-frequency EQ to 6dB of cut on the acoustic guitar channel, and sweep the frequencies between 60Hz and 1KHz.

☐ Is there a spot where the conflict goes away?

☐ If so, decrease the cut just to the point where you can hear the frequency conflict, then back it off a bit.

☐ Can you hear both instruments distinctly?

☐ Do they both sound acceptable?

E) If the conflict is in the higher frequencies, set a mid-frequency EQ to 6dB of cut, and sweep the frequencies between 1k and 8kHz.

☐ Is there a spot where the conflict goes away?

☐ If so, decrease the cut just to the point where you can hear the frequency conflict, then back it off a bit.

☐ Can you hear both channels distinctly?

☐ Do they both sound acceptable?

F) Unsolo both channels and listen with the rest of the track.

☐ Can you hear both tracks distinctly?

☐ Is there any other conflicts?

☐ If so, solo the tracks that are conflicting and repeat steps A through D.

Equalizing The Piano

The grand piano is an interesting instrument because it's so percussive and has such a wide frequency range. It can also play just about any role in an arrangement, from blending into the rhythm section, to being the pad element when played with long sustained chords or notes, to being the lead or fill element. This and the fact that the way a piano is miked means so much in its final recorded sound and you can see why it can be equalized so many different ways (see Figure 8.3).

TIP: The piano generally sounds full at around 80Hz and has presence at 3k to 5kHz, although too much boost in this frequency range can give it a honky-tonk quality.

Feature Frequencies

- 80Hz....................fullness

- 3kHz - 5kHz ....presence

Figure 8.3: Typical Piano Miking

Equalizing The Organ

As stated in Chapter 4, the organ is the quintessential instrument for the pad element of an arrangement. As a result, it's usually used as sort of a "glue" in the track and isn't always heard as a distinct instrument. It does have a wide frequency range though, and can have a huge low end, so care must be taken so it doesn't get in the way of the bass guitar or kick drum (see Figure 8.4).

Feature Frequencies

- 80Hz....................fullness

- 240Hz ....body

- 2kHz - 5kHz ....presence

- 100Hz...................HPF below

Figure 8.4: A Hammond A100 with a Leslie Speaker, which gives the organ its big low end

Equalizing Strings

Whether real or artificial, strings are frequently used as the finishing touch to an arrangement and frequently used as the pad element, although they tend to stick out of a mix more than an organ because of their mostly high frequency content.

Feature Frequencies

- 240Hz...................fullness
- 4k - 5kHz ....harshness
- 5kHz.....................shrill
- 7k - 10kHz ....scratchy

Equalizing Horns

Either brass or woodwinds contain mostly mid-range frequency content so they can easily conflict with guitars, vocals and snare drums/claps. While fullness comes at 120Hz, they can sound very piercing at 5kHz (especially brass instruments). Once again, an HPF can greatly clean up the sound.

Feature Frequencies

- 120Hz...................fullness
- 5kHz ....piercing
- 100Hz..................HPF below

Equalizing Percussion

Percussion can be categorized into two groups, the low frequency drum instruments like bongos, congas, djembe (see Figure 8.5), and udu; and high frequency instruments like shakers, tambourines and triangles. While each of these instruments have different frequency points, they can also easily conflict with other instruments in the mix.

Figure 8.5: A Djembe Drum

Hand percussion can easily conflict with the drum kit, guitars, vocals, high-hat, and strings in a race for frequency space. While they're very important to a mix because of the musical motion they convey, care must be taken when EQing.

TIP: *Generally speaking, the ring that bongos and congas sometimes have can be accentuated at 200Hz, while the slap comes at 5kHz. For high frequency percussion like shakers, 5k to 8kHz usually will make them more present. There's not normally much energy below 500Hz to 1kHz, so using the HPF below this point won't effect the sound at all while cleaning up some unwanted artifacts captured during recording.*

Feature Frequencies (Low Percussion)

- 200Hz...................ring

- 5kHz ....slap

Feature Frequencies (High Percussion)

- 500 - 1kHz............HPF below

- 5kHz - 10kHz........definition

The Principles of Equalization

Now that you've looked at the various techniques of equalization, here are some general equalization principles to keep in mind that can speed up the EQ process and keep you from chasing your EQ tail.

- The fewer instruments that are in the mix, the bigger each one should sound.

- Conversely, the more instruments in the mix, the smaller each one needs to be to fit them all in the mix.

- If it sounds muddy, cut some at 250Hz.

- If it sounds honky, cut some at 500Hz.

- Cut if you're trying to make things sound clearer.

- Boost if you're trying to make things sound different.

- You can't boost something that's not there in the first place.

- Use a narrow Q or bandwidth when cutting (like somewhere between 6 and 10), and a wide Q (like between 0.5 and 2) when boosting.

- If you want something to stick out, roll off the bottom; if you want it to blend in, roll off the top.

- Sometimes turning the level up on the mix channel works better than EQing.

Chapter 9
Adding Reverb

Like with many other aspects of mixing, the use of reverb is frequently misunderstood. Reverb is added to a track to create width and depth, but also to dress up an otherwise boring sound. The real secret is how much to use and how to adjust its various parameters to make it fit in the mix.

Reverb Basics

Before we get into adding and adjusting the reverb in your mix, let's look at some of the reasons to add reverb first.

When you get right down to it, there are four reasons to add reverb.

1. To make the recorded track sound like it's in a specific acoustic environment. Many times a track is recorded in an acoustic space that doesn't fit the song or the final vision of the mixer, producer or artist. You may record in a small dead room but want it to sound like it was in a large studio, a small reflective drum room, or a live and reflective church. A modern reverb plugin can take you to each of those environments and many more.

2. To add some personality and excitement to a recorded sound. Picture reverb as makeup on a model. She may look rather plain or even only mildly attractive until the makeup makes her gorgeous by covering her blemishes, highlighting her eyes, and accentuating her lips and cheekbones. Reverb does the same thing with some tracks. It can make the blemishes less noticeable, change the texture of the sound itself, or highlight it in a new way.

3. To make a track sound bigger or wider than it really is. Anything that's recorded in stereo naturally sounds bigger and wider than something recorded in mono because some of the natural ambience of the recording environment is captured as well. Most instrument or vocal recordings are recorded in mono though. As a result, an artificial space around the track may be needed.

TIP: Usually, reverb that has a short decay time (less than one second) will make a track sound bigger. Reverb with a longer decay time can push it back in the mix.

---

4. To move a track further back in the mix. While panning takes you from left to right in the stereo spectrum, reverb will take you from front to rear (see Figure 9.1). An easy way to understand how this works is to picture a band on stage. If you want the singer to sound like he or she's in front of the drum kit, you would add some reverb to the kit and keep the singer dry. If you wanted the horn section to sound like it was placed behind the kit, you'd add more reverb. If you wanted the singer to sound like he or she's in between the drums and the horns, you'd leave the drums dry and add a touch of reverb to the vocal, but less than the horns.

Figure 9.1: Reverb Puts The Sound From Front To Rear

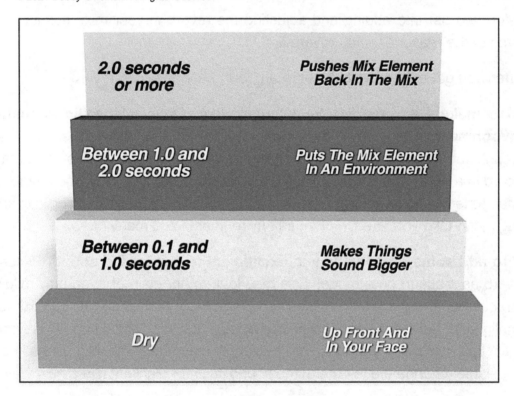

If we were going to get more sophisticated with this kind of layering, we'd use different reverbs for each of the instruments and tailor the parameters to best fit the sound we're going after.

Typical Reverb Parameters

One of the mistakes that many beginning mixers make is to simply add the reverb and never attempt to adjust it to make it work better with the track. While some sophisticated reverbs have exotic parameters like *Spin* and *Dampening* that can confuse all but the most experienced engineers, if you concentrate on just a few of the parameters that make the biggest difference in the sound and that are commonly found on most reverbs, you'll find that adjusting it gets a lot easier (see Figure 9.2).

Figure 9.2: Typical Reverb Parameter Controls
Courtesy of Valhalla DSP

The Major Parameter Controls

Although many sophisticated reverbs have a wide variety of somewhat obscure parameters, you can get exactly what you need for any mix with only the following five:

Reverb Type - Most variable reverb plugins have four environments that you can dial in; hall, room, chamber and plate (spring is another type that's not always included).

- A hall is a large space that has a long decay time and lots of reflections to it.

- A room is a much smaller space that can be dead or reflective, but it has a short decay time of about 1.5 seconds or less.

- An acoustic chamber is a tiled room with a speaker and microphone inside that many of the large studios used to specially build to create a great reverb sound (see Figure 9.3).

- A plate is a 4 foot hanging piece of metal with transducers on it that many studios used for artificial reverb when they couldn't afford to build a chamber (see Figure 9.4).

Figure 9.3: An Example Of An Acoustic Chamber

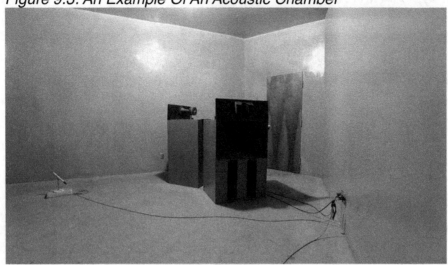

Figure 9.4: An EMT 140 Plate Reverb

Each of these reverb types have a distinctly different sound and there's no rule on which one to use. In most reverb plugins available today, you can usually just find one that you like and adjust the parameters to fit almost any song, or you can be more adventurous and try a different setting for every song. Either works.

Decay Time - The decay time represents how long it takes the reverb tail to fall off to where we can no longer hear it. Longer decay times push a track further back in the mix, while short ones (under about one second) make it sound bigger.

Pre-Delay - This parameter delays the entrance of the reverb. A pre-delay not only makes the reverb sound larger, but it keeps the reverb from clouding up the mix by allowing the listener to hear the attack of the mix element clearly before the reverb kicks in. In the days before electronic reverbs, pre-delay was achieved by using a tape recorder and setting the delay by the playback speed. Pre-delay can be usually be adjusted from about 50 to about 150 milliseconds or more (it depended on the make and model of the tape machine), which is about an eighth of a second.

TIP: *When in doubt, start with 20 milliseconds of pre-delay.*

High and Low-Pass Filters - The High and Low Pass filters are often overlooked, but they're the way that the tone of the reverb is shaped in order for the reverb to fit better in the mix. If you want to clearly hear the reverb, you typically wouldn't roll off the high-end much so it can stick out of the mix a bit. On the other hand, if you wanted the reverb to blend into the mix better, you'd use the LPF to roll-off the highs anywhere from 10k to as low as 2kHz.

Many times too much low end on the reverb just muddies up the mix. That's why you'd use the HPF to roll it off anywhere from 50Hz to as high as 600Hz. As an example, the famous Abbey Road Studios reverbs that have been heard on hundreds of hit records over the last 60 plus years (including every Beatle record), set their LPF at 10kHz and their HPF at 600Hz.

If a reverb that you're using doesn't have it's own built-in filters, you can always insert an equalizer or filter into the send or the return channel and roll the frequencies off there, or even boost them if needed.

Dry/Wet - The *Dry/Wet* control (sometimes called *Mix*) allows you to mix the reverb signal with the dry signal. This is essential for dialing in the correct amount of reverb if the plugin is inserted directly on a channel, but it's normally set to 100% wet when inserted into a dedicated effects return channel.

Many reverbs also include a parameter known as *Diffusion*, which simulates how reflective the walls in a particular space are. For instance, a more diffuse environment has hard walls with a lot of reflections, while one with softer walls has fewer reflections.

TIP: *A simple way to think of it is that high diffusion provides a thicker sounding reverb, while low diffusion is thinner sounding.*

Timing A Reverb To The Track

One of the secrets of hit making engineers is that they time the reverb to the track. That means timing both the pre-delay and the decay so it breathes with the pulse of the track. Here's how it's done.

Timing The Decay

The decay of a reverb is timed to the track by triggering it off of a snare or clap hit and adjusting the decay parameter so that the decay just dies by the next snare hit. The idea is to make the decay "breathe" with the track.

▶ *Exercise 9.1*: *Timing Reverb Decay*

All exercises in the chapter will use the example song "It's About Time," but they could also apply to any other song that you want to work on.

Before you begin any of the exercises in this chapter:

- Have two reverbs with the sends and returns already set up.

- Set one reverb to "Room" (label it Room Reverb) and the other to "Hall" (label it Hall Reverb). Refer to your DAW or console manual on how to do this.

- Solo the reverb returns or put them into *Solo Safe* (refer to your DAW or console manual on how to do this).

- Be sure that the *Dry/Wet* control is set to 100% wet.

A) Solo the Snare channel (or snare mix element), then raise the level of the send to the Room reverb until the reverb can be clearly heard.

☐ Does the snare sound distant?

☐ Does it sound bigger than before?

B) Adjust the *Decay* parameter of the reverb plugin until the reverb dies out before the next snare hit in the song (ghost notes don't count - just the hits on beats 2 and 4).

☐ Does the snare sound clearer?

C) If the Room reverb has additional parameters (i.e. Room 1 and Room 2, or Small, Medium, Large) try each parameter and set the *Decay* parameter as above.

☐ Does one setting sound better to you than another?

☐ Does the snare sound clearer?

D) Mute the send to the Room reverb and raise the level to the Hall reverb.

☐ Does the snare sound distant?

☐ Does it sound bigger than before?

☐ Does it sound bigger than the Room reverb?

E) If the Hall reverb has additional parameters (i.e. Room 1 and Room 2, or Small, Medium, Large) try each parameter and set the *Decay* parameter as above.

☐ Does one setting sound better to you than another?

☐ Does the snare sound clearer?

F) Adjust the *Decay* parameter until the reverb dies out before the next snare hit in the song.

☐ Does the snare sound clearer?

☐ Does it sound bigger?

E) Unsolo the Snare mix element and listen in the track.

☐ Does the snare sound clear or muddy?

☐ Does it push it back in the track

☐ Does it sound bigger?

Timing The Pe-Delay

Timing the pre-delay to the tempo of the track can add depth without the reverb being noticeable. You can time the pre-delay to the track by using the *Sync* option on the reverb plugin so it automatically syncs to the tempo of the track, or by tapping the time in if your plugin allows you to do so.

You can also determine the time manually using the following formula:

7,500 ÷ the beats per minute of the track = delay time in milliseconds

As an example:

7,500 ÷ 125 BPM = 60 milliseconds

This is the delay of a thirty-second note. If that's too long, you can divide the result of the formula (60 milliseconds) by 2 to get a 1/64th note delay of 30 milliseconds. Or double it for a delay of 1/16th note at 120 milliseconds. Any other amount that's divisible, like 45 ms or 90ms, will also sound pretty good.

Another way to time the pre-delay to the track is the use one of the many smartphone applications available.

▶ *Exercise 9.2*: *Timing Reverb Pre-Delay*

Solo the Snare Mix element again, as well as the reverb returns (or put them into *Solo Safe* - refer to you DAW or console manual on how to do this).

A) Find the BPM of the song by using one of the following:
- the tap feature found on most DAWs

- the tap function of a smartphone app.

- If the reverb plugin has a *Sync* function to automatically sync the reverb to the tempo of the track, use that instead.

B) To manually find the pre-delay time, use the formula found above (7,500 ÷ song BPM) to find the pre-delay time (using the *Sync* function will automatically find a pre-delay time).

C) Set the *Pre-delay* parameter of the Room reverb to the time you determined in step B.

☐ Does the snare sound bigger?

☐ Does it sound more distinct?

☐ Do you have to adjust the level of the reverb?

☐ Can you hear the attack of the snare element better?

D) Unsolo the Snare channel and the reverb returns.

☐ Does the snare sound bigger when mixed in the track?

☐ Does it sound more distinct?

☐ Do you have to adjust the level of the reverb?

☐ Can you hear the attack better?

E) Divide the reverb time by two and adjust the Pre-Delay parameter to this number (i.e. 20ms).

☐ Does the snare sound bigger in the track?

☐ Does it sound more distinct?

☐ Do you have to adjust the level of the reverb?

☐ Can you hear the attack better?

F) Take your original pre-delay time and multiply it by two (i.e. 80ms).

☐ Does the snare mix element sound bigger in the track?

☐ Can you hear it better in the mix with a longer pre-delay?

☐ Do you have to adjust the level of the reverb?

☐ Can you hear the attack of the snare mix element better?

Reverb Setup

As we spoke about in Chapter 2, setting up your reverbs before you begin to mix can be a huge time saver. You'll have to make a few final tweaks as you go along, but that happens with just about every parameter of a mix anyway. Here's a quick setup when you don't have time to fine tune the reverb settings.

The Two Reverb Quick Setup Method

This setup is designed to get you up and running quickly, with the parameters in a general position to where they almost always sound at least acceptable, and sometimes even surprisingly good (see Figure 9.5). This method works well when you're tracking and need some quick reverb, or for doing rough mixes when you don't have time for a more complete setup. It uses two different reverbs.

Figure 9.5: The Two Reverb Setup

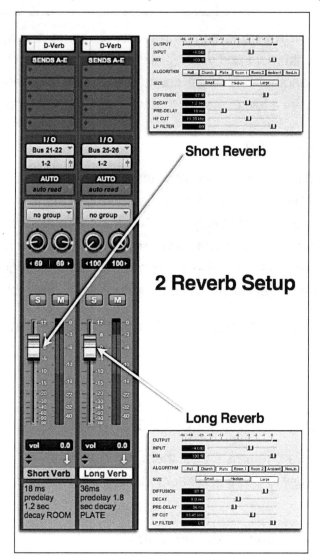

You can start with the following parameters:

- **Reverb One** - This reverb will be used just for the drums. Set it to a *Room* with the decay at 1.0 second and a pre-delay of 20 milliseconds. If a LPF is available, set it to 10kHz or even 8kHz. If a HPF is available, set it to 100Hz.

- **Reverb Two** - This reverb is meant for all other instruments and vocals. Set it to a *Plate* with a 1.8 second decay time and a pre-delay of 20 milliseconds. If a LPF is available, set it to 10kHz or even 8kHz. If a HPF is available, set it to 100Hz.

It's surprising how good everything can sound with this quick setup, but *ideally you'll be setting up the pre-delay and decay time on your reverb plugins as in the Exercises 9.1 and 9.2.*

Adding Reverb To Instruments

There's no rule for where and how reverb is added to a mix. In some mixes, a single reverb can work for every instrument and vocal (just like in all those great hits of the 50s, 60s and 70s), while another mix might sound better with a separate reverb for every mix element. Let's try a few exercises that will use each scenario.

Adding Reverb To The Drum Kit

The drum kit is interesting and different from the other instruments in the mix in that it may use several different reverbs in order for it to lay correctly in the track. For instance, the kick drum may be left completely dry, while the snare drum may have its own separate reverb, or an additional one from what's applied to the rest of the kit. In fact, the cymbals and hat may also have more of a room sound, while the toms may get one with a longer decay and put further back in the mix.

Then on the other hand, a single reverb may be just enough to make the kit sound great.

▶ ***Exercise 9.3****: Adding Reverb To The Snare Drum*

Solo the entire drum kit and the reverb returns (or put them into *Solo Safe* - refer to your DAW or console manual on how to do this). *Be sure that the Dry/Wet control is set to 100% wet..*

A) As you raise the reverb amount, ask yourself:

☐ What kind of sound are you looking for?

☐ Do you want the snare to sound bigger or do you want to push it back in the mix?

B) Increase the level of the send to the Room reverb on the Snare channel until the reverb can be clearly heard.

☐ Does it sound distant?

☐ Does it sound bigger than before?

☐ Does it seem to move behind the other mix elements?

☐ Does it sound better with no reverb at all?

☐ Does it fit better when you unsolo the snare so you can hear it in the mix?

☐ Can you hear the difference between the snare with and without reverb?

C) Now send snare to the reverb set to Hall.

☐ Does it sound better?

☐ Does it sound distant?

☐ Does it sound bigger than before?

☐ Does it seem to move behind the other mix elements?

☐ Does it sound better with no reverb at all?

☐ Does it fit better when you unsolo the snare so you can hear it in the mix?

D) Now send it to both reverbs.

☐ Does the it sound distant?

☐ Does it sound bigger than before?

☐ Does it seem to move behind the other instruments?

☐ Does it sound better with no reverb at all?

☐ Does it fit better when you unsolo the snare so you can hear it in the mix?

E) What happens if you increase or decrease the *Pre-delay* setting as in Exercise 9.2?

☐ Does it sound distant?

☐ Does it sound bigger than before?

☐ Can you hear it better with a longer pre-delay?

☐ Does it seem to move behind the other mix elements?

☐ Does it sound better with no reverb at all?

☐ Does it fit better when you unsolo the snare so you can hear it in the mix?

F) What does it sound like if you increase or decrease the *Decay* setting as in Exercise 9.1?

☐ Does it sound distant?

☐ Does it sound bigger than before?

☐ Does it seem to move behind the other instruments?

☐ Does it sound better with no reverb at all?

☐ Does it fit better when you unsolo the snare so you can hear it in the mix?

G) After choosing the reverb and its settings, listen to the snare with the rest of the mix.

☐ Add enough reverb so that you can just hear it, if any is needed. You'll dial in the exact amount needed later after reverb is added to the rest of the mix elements.

► Exercise 9.4: Adding Reverb To The Kick Drum

A) Visualize the mix.

☐ What kind of sound are you looking for?

☐ Do you want the Kick to sound bigger or do you want to push it back in the mix?

B) Solo the entire drum kit, then on the Kick channel, increase the level of the send to the Room reverb until the reverb can be clearly heard.

☐ Does it sound distant?

☐ Does it sound bigger than before?

☐ Does it seem to move behind the other instruments?

☐ Did it change the sound of the snare drum?

☐ Does it sound better with no reverb at all?

☐ Does it fit better when you unsolo the kick so you can hear it in the mix?

C) Now send Kick to reverb set to Hall.

☐ Does it sound better?

☐ Does it sound distant?

☐ Does it sound bigger than before?

☐ Does it seem to move behind the other instruments?

☐ Does it sound better with no reverb at all?

☐ Does it fit better when you unsolo the kick so you can hear it in the mix?

D) Now send it to both reverbs.

☐ Does the it sound distant?

☐ Does it sound bigger than before?

☐ Does it seem to move behind the other instruments?

☐ Does it sound better with no reverb at all?

☐ Does it fit better when you unsolo the kick so you can hear it in the mix?

E) What happens if you increase or decrease the *Pre-delay* setting?

☐ Does it sound distant?

☐ Does it sound bigger than before?

☐ Can you hear it better with a longer pre-delay?

☐ Does it seem to move behind the other instruments?

☐ Does it sound better with no reverb at all?

☐ Does it fit better when you unsolo the Kick so you can hear it in the mix?

F) What does it sound like if you increase or decrease the *Decay* setting?

☐ Does it sound distant?

☐ Does it sound bigger than before?

☐ Does it seem to move behind the other instruments?

☐ Does it sound better with no reverb at all?

☐ Does it fit better when you unsolo the Kick so you can hear it in the mix?

G) Does the Kick have more power with or without the reverb?

_____

▶ *Exercise 9.5: Adding Reverb To The Toms*

A) Solo the drum kit, the visualize the mix.

☐ What kind of sound are you looking for?

☐ Do you want the Toms to sound bigger or do you want to push them back in the mix?

B) Increase the level of the send to the Room reverb on the Toms channels until the reverb can be clearly heard.

☐ Do they sound distant?

☐ Do they sound bigger than before?

☐ Do they seem to move behind the other instruments?

☐ Did it change the sound of the other drums?

☐ Do they sound better with no reverb at all?

☐ Do they fit better when you unsolo the Toms so you can hear them in the mix?

C) Now send toms to reverb set to Hall.

☐ Do they sound better?

☐ Do they sound distant?

☐ Do they sound bigger than before?

☐ Do they seem to move behind the other instruments?

☐ Do they sound better with no reverb at all?

☐ Do they fit better when you unsolo the Toms so you can hear them in the mix?

D) Now send it to both reverbs.

☐ Do they it sound distant?

☐ Do they sound bigger than before?

☐ Do they seem to move behind the other instruments?

☐ Do they sound better with no reverb at all?

☐ Do they fit better when you unsolo the Toms so you can hear them in the mix?

E) What happens if you increase or decrease the *Pre-delay* setting?

☐ Do they sound distant?

☐ Do they sound bigger than before?

☐ Can you hear the toms better with a longer pre-delay?

☐ Do they seem to move behind the other instruments?

☐ Do they t sound better with no reverb at all?

☐ Do they fit better when you unsolo the Toms so you can hear them in the mix?

F) What does it sound like if you increase or decrease the *Decay* setting?

☐ Do they sound distant?

☐ Do they sound bigger than before?

☐ Do they seem to move behind the other instruments?

☐ Do they sound better with no reverb at all?

☐ Do they fit better when you unsolo the Toms so you can hear them in the mix?

G) After choosing the reverb and its settings, listen to the toms with the rest of the mix.

☐ Add enough reverb so that you can just hear it, if any is needed. You'll dial in the correct amount later after the reverb is added to the rest of the mix elements.

> **Exercise 9.6:** *Adding Reverb To The Cymbals.*

A) Visualize the mix.

☐ What kind of sound are you looking for?

☐ Do you want the Cymbals to sound bigger or do you want to push it back in the mix?

B) Increase the level of the send to the Room reverb until the reverb can be clearly heard.

☐ Does it sound distant?

☐ Does it sound bigger than before?

☐ Does it seem to move behind the other instruments?

☐ Did it change the sound of the other drums?

☐ Does it sound better with no reverb at all?

☐ Does it fit better when you unsolo the Cymbals so you can hear them in the mix?

C) Now send Cymbals to reverb set to Hall.

☐ Does it sound better?

☐ Does it sound distant?

☐ Does it sound bigger than before?

☐ Does it seem to move behind the other instruments?

☐ Does it sound better with no reverb at all?

☐ Does it fit better when you unsolo the Cymbals so you can hear them in the mix?

D) Now send it to both reverbs.

☐ Does the it sound distant?

☐ Does it sound bigger than before?

☐ Does it seem to move behind the other instruments?

☐ Does it sound better with no reverb at all?

☐ Does it fit better when you unsolo the Cymbals so you can hear them in the mix?

E) What happens if you increase or decrease the *Pre-delay* setting?

☐ Does it sound distant?

☐ Does it sound bigger than before?

☐ Can you hear them better with a longer pre-delay?

☐ Does it seem to move behind the other instruments?

☐ Does it sound better with no reverb at all?

☐ Does it fit better when you unsolo the Cymbals so you can hear them in the mix?

F) What does it sound like if you increase or decrease the *Decay* setting?

☐ Does it sound distant?

☐ Does it sound bigger than before?

☐ Does it seem to move behind the other instruments?

☐ Does it sound better with no reverb at all?

☐ Does it fit better when you unsolo the Cymbals so you can hear them in the mix?

G) After choosing the reverb and its settings, listen to the cymbals with the rest of the mix.

☐ Add enough reverb so that you can just hear it, if any is needed. You'll dial in the correct amount later after the reverb is added to the rest of the mix elements.

▶ Exercise 9.7: *Adding Reverb To The High-Hat*

A) Visualize the mix.

☐ What kind of sound are you looking for?

☐ Do you want the Hat to sound bigger or do you want to push it back in the mix?

B) Proceed as in the previous exercises.

Adding Reverb To The Bass

Since the bass is so important to the power of the song, and because it has an abundance of low frequency information, reverb is rarely added. That doesn't mean there isn't the occasion where it's effective though ("For The Love Of Money," the 70s hit by the O'Jays, comes to mind). Sometimes a very little bit of a very short room reverb can make the bass sound a bit bigger, or make it sound like the direct bass recording was in the same room when recording the drums. Either way, it's always worth a try to see if a little reverb just might add some magic.

▶ Exercise 9.8: *Adding Reverb To The Bass*

A) Solo the bass and the reverb returns. Make sure the *Decay* and *Pre-delay* are both timed to the track.

☐ What kind of sound are you looking for?

☐ Do you want the bass to sound bigger or do you want to push it back in the mix?

B) Increase the level of the send to the Room reverb until the reverb can be clearly heard.

☐ Does the bass sound distant?

☐ Does it sound bigger than before?

☐ Does it sound better with no reverb at all?

☐ What does it sound like when you play unsolo the bass so you can hear it with the track?

C) Mute the Room reverb and increase the send level to the Hall reverb.

☐ Does the bass sound better than the Room reverb?

☐ Does the it sound distant?

☐ Does it sound bigger than before?

D) What does the bass sound like with both reverbs added?

E) Increase the *Pre-delay* time setting on the reverb you've chosen.

☐ Does the bass sound distant?

☐ Does it sound bigger than before?

☐ Can you hear it better with a longer pre-delay?

☐ What does it sound like if you decrease the *Pre-delay* setting?

F) What does the bass sound like if you increase or decrease the *Decay* setting?

G) Does the bass fit any better in the mix if you increase the HPF frequency on the reverb?

H) Does it fit any better in the mix if you lower the LPF frequency on the reverb?

I) Does the bass lose power when the reverb is added?

☐ Does it distract from the other mix elements?

_____

▶ *Exercise 9.9: Tweaking The Sound Of The Reverb*

Sometimes the reverb just muddies up the track because of too much low end and too much high end. We can tailor the sound of the reverb by just adding high-pass and low-pass filters.

B) To set up the filter do the following:

- **Insert an EQ with both a high-pass and low-pass filter on the Hall Reverb Aux channel. *Place it before the reverb plugin.***

- ***Solo the lead Vocal, insert an aux send to Hall Reverb, and raise the level until you can hear the reverb.***

B) Set the Q of the HPF to 12dB/oct and the frequency to 100Hz.

☐ Can you hear the difference when you bypass the HPF?

☐ Did the reverb get bigger or smaller sounding?

☐ Did you have to adjust the reverb level?

☐ Is the reverb muddy or distinct?

C) Now set the HPF to 250Hz.

☐ Can you hear the difference when you bypass the HPF?

☐ Did the reverb get bigger or smaller sounding?

☐ Did you have to adjust the reverb level?

☐ Is the reverb muddy or distinct?

D) Now set the HPF to 500Hz?

☐ Can you hear the difference when you bypass the HPF?

☐ Did the reverb get bigger or smaller sounding?

☐ Did you have to adjust the reverb level?

☐ Is the reverb muddy or distinct?

E) Set the Q of the LPF to 12dB/oct and the frequency to 8kHz.

☐ Can you hear the difference when you bypass the LPF?

☐ Did the reverb get bigger or smaller sounding?

☐ Did you have to adjust the reverb level?

☐ Is the reverb muddy or distinct?

F) Now set the LPF to 4kHz.

☐ Can you hear the difference when you bypass the LPF?

☐ Did the reverb get bigger or smaller sounding?

☐ Did you have to adjust the reverb level?

☐ Is the reverb muddy or distinct?

G) Now set the HPF to 2kHz.

☐ Can you hear the difference when you bypass the LPF?

☐ Did the reverb get bigger or smaller sounding?

☐ Did you have to adjust the reverb level?

☐ Is the reverb muddy or distinct?

H) Choose the filter settings that you think work best in the mix.

☐ Can you hear the difference when you bypass the EQ?

☐ Did you have to adjust the reverb level?

I) Unsolo the Vocal and listen to the mix. Readjust the reverb level so you can hear it clearly in the mix.

- ☐ Can you hear the difference when you bypass the EQ?
- ☐ Can you hear the difference when you mute the aux send so there's no reverb?

Adding Reverb To The Vocal

Since the lead vocal is usually the focal point of the song, the reverb setting is critical because of how it can make the vocal fit in the mix. Pick the right one and it'll add that extra professional-sounding sheen that all hit records have. Pick the wrong one and it'll sound washed out and lost in the track, or you won't hear it at all.

Many times a lead vocal has a lot more reverb on it than it seems, but it's disguised by the way its bandwidth is tailored by the HPF and LPF. Other times, it's important to hear the reverb and every effort is made to maintain or even increase its high frequency response. Ballads that have a long period of space in between vocal lines will usually benefit from a longer reverb decay that's obvious.

In the end, you're usually trying to put the vocalist in a space and not push her back in the mix, which is the opposite of what you're trying to do with background vocals.

Background vocals sometimes are just put in a space, pushed back in the mix from the lead vocal, or even made bigger than life thanks to a very short (0.2 to 0.4 second decay time) reverb. If the background vocals are singing harmony with the lead vocal, they sometimes need to have the same reverb as the lead vocal, but most of the time you want them to be distinguished separately, so a different reverb is used.

▶ *Exercise 9.10*: *Adding Reverb To The Lead Vocal*

A) Solo the lead Vocal and the reverb returns. Make sure the *Decay* and *Pre-delay* are both timed to the track.

- ☐ What kind of sound are you looking for?
- ☐ Do you want to put the lead vocal in a space or do you want to push it back in the mix?

B) Increase the level of the send to the Room reverb until the reverb can be clearly heard.

- ☐ Does it put the vocalist in another environment?
- ☐ Does the the vocal sound distant?
- ☐ Does it sound better with no reverb at all?
- ☐ What does it sound like when you play unsolo the vocal so you can hear it with the track?

C) Mute the Room reverb and increase the send level to the Hall reverb.

☐ Does the lead vocal sound better with the Hall than the Room reverb?

☐ Does the it sound distant?

☐ Does the artificial space better fit the track?

D) What does the lead vocal sound like with both reverbs?

E) Increase the *Pre-delay* setting on the whatever reverb you've chosen.

☐ Can you distinguish the vocal better?

☐ What does it sound like if you lower the *Pre-delay* setting?

☐ Can you hear it better with a longer pre-delay?

☐ Is the vocal muddy sounding?

F) What does the lead vocal sound like if you increase or decrease the Decay setting?

☐ Can you hear the vocal more or less clearly?

G) Does the vocal fit any better in the mix if you raise the High-Pass Filter frequency on the reverb?

H) Does it fit any better in the mix if you lower the Low-Pass Filter frequency on the reverb?

I) After choosing the reverb and its settings, listen to the lead vocal with the rest of the mix. Add enough reverb so that you can just hear it if any is needed. You'll dial in the correct amount later after the reverb is added to the rest of the instruments.

▶ *Exercise 9.11: Adding Reverb To The Background Vocals*

A) Solo the background vocals and the reverb returns. Make sure the *Decay* and *Pre-delay* are both timed to the track.

☐ Proceed as in the previous exercise.

Adding Reverb To Guitars

While guitars are percussive by nature, they also can have a sustaining quality. In fact, an electric guitar playing power chords can be a very effective pad element in a song, and pads generally like longer reverb decays. On the other hand, many guitar parts sound much more interesting when made to sound larger than life with a very short (0.2 or 0.3 seconds decay time) reverb. Whichever you choose, you'll find that a timed reverb almost always works with just about any kind of guitar.

▶ *Exercise 9.12*: Adding Reverb To The Guitar

C) As in the previous exercises, while listening to the full mix, add reverb until you can just hear it in the track.

☐ Does the mix element blend with the track better?

☐ Does the reverb make it sound more polished?

B) Now mute the reverb.

☐ Can you hear a difference?

☐ Does the reverb work for the mix element or not?

C) Add another aux channel and insert a new reverb just for the guitars as in the previous exercises.

☐ Does it sound better with a different reverb plugin?

☐ Does it sound better with shorter or longer settings?

☐ Does it sound better with different filter settings?

☐ Does it sound better untimed?

Adding Reverb To Keyboards

Since there is such a wide variety of keyboards, they each demand their own approach when it comes to adding reverb. For instance, since a piano is so percussive, reverbs that are timed to the track usually fit better than reverbs that are not. The exception might be in a solo piano or classical situation, where you're more concerned with putting the piano in an artificial space or enhancing any ambience already there instead of worrying about how it fits with a rhythm section.

An organ, on the other hand, plays more long sustaining chords, so while it might sound pretty good with a timed reverb, that's not always essential. You usually don't try to make it sound larger than life, so simply putting it in an artificial space works well.

The synthesizer is a hybrid of both however. It can have all the attack and percussion of a piano or the sustaining pad element of an organ, and anywhere in between. You may want it to sound larger than life, push it back in the mix, or just put it in an artificial space.

▶ *Exercise 9.13*: Adding Reverb To The Piano

D)As in the previous exercises, while listening to the full mix, add reverb until you can just hear it in the track.

☐ Does the mix element blend with the track better?

☐ Does the reverb make it sound more polished?

B) Now mute the reverb.

☐ Can you hear a difference?

☐ Does the reverb work for the mix element or not?

C) Add another aux channel and insert a new reverb just for the piano as in the previous exercises.

☐ Does it sound better with a different reverb plugin?

☐ Does it sound better with shorter or longer settings?

☐ Does it sound better with different filter settings?

☐ Does it sound better untimed?

☐ Does it sound better if the piano has more reverb and sits further back in the mix, or less reverb and sits in the front?

▶ Exercise 9.14: *Adding Reverb To The Organ*

E)As in the previous exercises, while listening to the full mix, add reverb until you can just hear it in the track.

☐ Does the mix element blend with the track better?

☐ Does the reverb make it sound more polished?

B) Now mute the reverb.

☐ Can you hear a difference?

☐ Does the reverb work for the mix element or not?

C) Add another aux channel and insert a new reverb just for the organ as in the previous exercises.

☐ Does it sound better with a different reverb plugin?

☐ Does it sound better with shorter or longer settings?

☐ Does it sound better with different filter settings?

☐ Does it sound better untimed?

☐ Does it sound better if the organ has more reverb and sits further back in the mix, or less reverb and sits in the front?

▶ Exercise 9.15: *Adding Reverb To Synthesizers*

F)As in the previous exercises, while listening to the full mix, add reverb until you can just hear it in the track.

☐ Does the mix element blend with the track better?

☐ Does the reverb make it sound more polished?

B) Now mute the reverb.

☐ Can you hear a difference?

☐ Does the reverb work for the mix element or not?

C) Add another aux channel and insert a new reverb just for the synthesizer as in the previous exercises.

☐ Does it sound better with a different reverb plugin?

☐ Does it sound better with shorter or longer settings?

☐ Does it sound better with different filter settings?

☐ Does it sound better untimed?

☐ Does it sound better if the synthesizer has more reverb and sits further back in the mix, or less reverb and sits in the front?

Adding Reverb To Strings

Real or virtual strings sound best when we place them in a medium to large artificial hall because that's how we're used to hearing them. Usually this means a Hall or Church setting (if the plugin has one), with a decay time of two seconds or more. While a short pre-delay of about 10 or 20ms might allow the attack of the strings to be more noticeable, longer pre-delays don't usually suit any instrument that's a pad element in the mix, like the strings.

TIP: Because of the long sustaining quality of the sound of a string section, the decay time is far less important than it is with other instruments.

▶ *Exercise 9.16: Adding Reverb To The String Section*

A) As in the previous exercises, while listening to the full mix, add reverb until you can just hear it in the track.

☐ Does the mix element blend with the track better?

☐ Does the reverb make it sound more polished?

B) Now mute the reverb.

☐ Can you hear a difference?

☐ Does the reverb work for the mix element or not?

C) Add another aux channel and insert a new reverb just for the strings as in the previous exercises.

☐ Does it sound better with a different reverb plugin?

☐ Does it sound better with shorter or longer settings?

☐ Does it sound better if the strings have more reverb and sit further back in the mix, or less reverb and sit in the front?

☐ Does it sound better if the reverb is timed or un-timed?

☐ Does it sound better with longer reverb decay times (beyond 2 seconds)?

☐ Does it sound better with different filter settings?

Adding Reverb To Horns

We're usually used to hearing a horn (whether it's brass or woodwind) or horn section in a space, but you also might want it further back in the mix, and even occasionally, larger than life. Because most horns have a brisk attack and release, a timed reverb works very well.

> ### ▶ Exercise 9.17: Adding Reverb To Horns

A) As in the previous exercises, while listening to the full mix, add reverb to the Horns until you can just hear it in the track.

☐ Does the mix element blend with the track better?

☐ Does the reverb make it sound more polished?

B) Now mute the reverb.

☐ Can you hear a difference?

☐ Does the reverb work for the mix element or not?

C) Add another aux channel and insert a new reverb just for the strings as in the previous exercises.

☐ Does it sound better with a different reverb plugin?

☐ Does it sound better with shorter or longer settings?

☐ Does it sound better if the strings have more reverb and sit further back in the mix, or less reverb and sit in the front?

☐ Does it sound better if the reverb is timed or un-timed?

☐ Does it sound better with longer reverb decay times (beyond 2 seconds)?

☐ Does it sound better with different filter settings?

Adding Reverb To Percussion

Just like drums, percussion is made up of short bursts of sound with strong attacks. This means that they benefit greatly from pre-delay and usually like reverbs that are timed to the track.

TIP: *Normally you're not trying to make any of the percussion instruments sound bigger or push them back in the track, you're trying to put them into an environment.*

> **Exercise 9.18**: Adding Reverb To Bongos And Congas

A) Follow steps as in the previous exercise, but keep in mind that faster decays and pre-delays sometimes work best.

> **Exercise 9.19**: Adding Reverb To Shakers And Triangles

A) Follow steps as in the previous exercise, but keep in mind that faster decays and pre-delays sometimes work best.

Layering The Mix

By now all the vocals and instruments should sound pretty good and everything should be in it's own artificial space. Now it's time to layer the mix.

When layering the mix, we'll be thinking of three things:

1. Are the instruments in front or behind each other in the pleasing manner in the mix

2. Does one of the mix elements need a completely different reverb sound, and therefore its own reverb.

3. Does a mix element need an effect other than reverb, like a delay or modulation (we'll cover these in the next two chapters).

When it comes to layering the mix, we're talking about the reverb balance of each instrument, loop, beat, or vocal in the mix. Some tracks will be up front and in your face and therefore won't have much reverb on them, or the reverb will be tailored so it isn't obvious using the high and low pass filters. Others will have more and more reverb on them and seem to be pushed back in the soundstage as a result. If we were to visually imagine what our mix would be like it would look like Figure 9.6.

Figure 9.6: A Visual Look At Layering A Mix

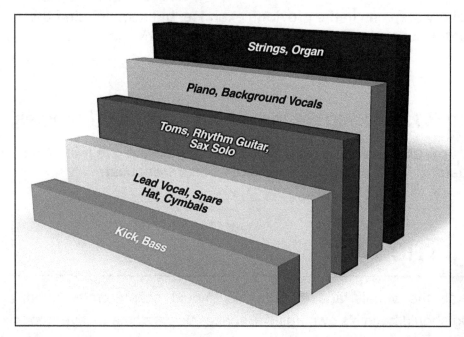

Using a combination of different sounding reverbs with different pre-delays and decay times, along with using simply more or less reverb, you can layer the mix like anything the pros might do, because that's how they do it. Keep in mind that each song is unique, so sometimes a song won't require much reverb or maybe even none at all, but it's all up to you and your ears how you apply what you've learned in this chapter.

▶ *Exercise 9.20: Layering The Mix*

A) Listen to the entire mix with all the reverbs that you've applied so far.

☐ Starting from the vocal (or the most important mix element), listen to each track and decide how close to the listener it should sound. Add more reverb if you want to push it back, and less if you want to bring it to the forefront.

☐ On the instruments that have the short reverbs, decide how big you want each track to sound. Add more reverb if you want it bigger, and less if you want it smaller. Remember that you may have to adjust the track levels as a result of increasing or decreasing the reverb.

☐ If there's a mix element that still doesn't fit into the mix, try a dedicated reverb with completely different settings from the others that you're using and follow steps A through I of exercise 9.12.

Chapter 10
Adding Delay

Delay (sometimes called Echo) is an integral part of a mixer's toolbox because it's able to make things sound larger than life or push them back in the mix, just like reverb. The advantage is that it does so by being somewhat less noticeable than reverb. In fact, there are some recordings where the only effect is delay and not a speck of reverb is used anywhere, which makes it very powerful.

Delay Basics

Many of the reasons for using delay are the same as with reverb. For instance, delay can be used to:

1. Push a sound back in the mix. By using a longer delay, the mix element will seem further away from the listener if the level is high enough. Adding more repeats also enhances this affect.

2. Make an instrument or vocal sound larger than life. A very short delay (under 40 milliseconds) reinforces the dry signal while artificially reproducing what's known as the "first reflection" in the room, which is the most powerful and audible part of natural room ambience. Also, by panning the delay to one side of the mix and the dry signal to the other, it widens the sound of the track in the stereo soundstage.

3. Add an artificial double. By adding a 50 to 100ms delay, you can artificially create the slap-back, double track effect heard on so many of the hits of the 50s (because that's the only effect they had back then).

4. Add a "glue" to the mix. A delay that's timed to the tempo of the song essentially disappears, but it has the effect of melding the track together in a way that reverb can't do. Veteran mixers call it the "glue" to the mix.

Typical Delay Parameters

There are fewer control parameters than with reverb, but that doesn't make a typical delay any less powerful (see Figure 10.1). They are:

Figure 10.1: Typical Delay Parameter Controls
Courtesy of Avid Technology Inc.

Delay Time - This is similar to decay time in reverb except it's usually measured in milliseconds since it's rare that it's ever as long as a second (which is 1,000 milliseconds).

Repeats - The number of repeats (sometimes called *Regeneration* or *Feedback*) ranges from 0 to infinity, but most of the time it's set for 0 to approximately 3 repeats. Too many repeats makes the mix muddy while too few makes the delay less obvious.

TIP: Setting the Repeats parameter too high will result in an endless loop of runaway feedback, which is highly undesirable.

Filters - Just like reverb, the sound of a delay can be affected greatly by high-pass and low-pass filters. On most delay plugins currently available these are built in, but in many cases you might have to insert them into the effects channel. Just like reverb, inserting a HPF can clean up a mix by eliminating the low frequency information, while eliminating the high frequency with a LPF can make the delay less obvious and have it blend into the mix more easily.

Dry/Wet - The *Dry/Wet* control (sometimes called *Mix*) allows you to mix the delayed signal with the dry signal. This is essential for dialing in the correct amount of delay if the delay plugin is inserted on an individual channel, but it's normally set to 100% wet when inserted into a dedicated effects return channel.

Sometimes delay plug-ins also have parameter controls like *Sync, Tempo* and *Meter*, which allow the delay to be easily timed to the tempo of the song.

The Haas Effect

The Haas Effect is a very useful psychoacoustic theory that states that a delay of 40 milliseconds or less (depending on which text book you read - sometimes it's noted as 30 milliseconds) is not perceived as a distinct event. That means that if a delay on a snare drum was set to 50 milliseconds, you would hear two separate events - the initial snare hit, then the delay. But if you set the delay to less than 30 milliseconds, the two would blend together and you'd hear them both as a single event (see Figure 10.2).

Figure 10.2: The Haas Effect

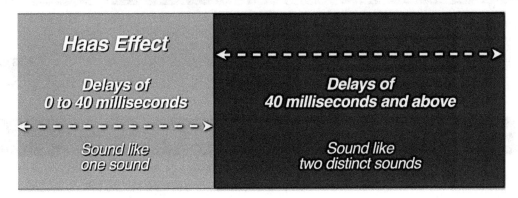

This is important because it means that a short delay of less than 30 milliseconds can be used to thicken a sound that seems a bit thin sounding, or it can be used to "stereoize" a mono track by panning the dry track to one side and the delayed to the other. Regardless how you use it, the Haas Effect is a very powerful, if sometimes overlooked tool in your mixing toolbox.

Timing The Delay To The Tempo Of The Song

Just like with reverb, it's important to time your delay to the tempo of the song unless you want it to be very obvious. The difference is that there are so many more possibilities, depending upon your needs. Timing the delay to the beat of the song can add depth without the delay being noticeable.

Many delay plug-ins allow you to sync to the BPM of the song, which automatically determines the delay times. If this option is not available, you can time the delay to the song by using the following formula:

60,000 ÷ by the beats per minute of the song = delay time in milliseconds

As an example:

60,000 ÷ 125 BPM = 480 milliseconds

This is the delay of a 1/4 note. If that's too long, you can divide the result of the formula (480 milliseconds) by 2 to get a 1/8th note delay of 240 milliseconds. Divide by 2 again and you get a 1/16th note delay of 120 milliseconds. You can keep dividing by 2 to get smaller and smaller note divisions.

Other Note Divisions

Sometimes the delay sounds better other note denominations such as triplets or dotted eights, sixteenths, etc are used. These denominations can be determined using the following formula:

Delay Time X 1.5 = Dotted Value

For example:

480ms (the quarter note 125 BPM delay from the previous example) x 1.5 = 720ms (Dotted Quarter Note)

Delay Time X .667 = Triplet Value

For example:

480ms (quarter note 125 BPM delay) x .667 = 320ms (Quarter Note Triplet)

As with the straight notes (quarter, eighths, etc.), you can continually divide the above values in half until you get the desired denomination.

TIP: While the straight note denominations of 1/4, 1/8th, 1/16th and on can provide depth to a mix, triplet and dotted note denominations are great for adding glue. They can make an element blend seamlessly in the mix.

▶ *Exercise 10.1: Setting A Haas Delay*

All exercises in the chapter will use the example song "It's About Time," but they could also apply to any other song that you want to work on.

Before you begin the following exercises:

- Insert a delay plugin on an aux channel and set the buss input (refer to your DAW or console manual on how to do this).

- Solo the delay returns or put them into *Solo Safe* (refer to your DAW or console manual on how to do this).

- Be sure that the *Dry/Wet* control is set to 100% wet.

A) Find the BPM of the song by using one of the following:

- the tap feature found on most DAWs

- the tap function of a smartphone app.

- If the delay plugin has a *Sync* function to automatically sync the delay to the tempo of the track, use that instead.

B) To manually find the delay time, use the formula found above (60,000 ÷ song BPM).

C) Set the time for whatever time denomination (1/16th, 1/32nd, 1/64th note) that sets the delay time below 40ms with no repeats (keep on dividing the delay time in half until you get below 40ms - preferably around 20ms).

D) Solo the Snare channel(s) (or snare mix element), then raise the level of the send to the delay until the delay can be clearly heard.

- ☐ Does the snare sound bigger?

- ☐ Does it sound more distinct?

- ☐ Does it sound thicker?

E) Solo the rest of the drums.

- ☐ Do you have to increase the level to hear the delay?

- ☐ Does the Snare sound bigger in the mix?

- ☐ Can you hear a difference when you mute the delay?

- ☐ Does it sound more distinct?

F) Set the Feedback to about 20%.

- ☐ Does it sound better or worse?

- ☐ Does it blend in with the mix or does it stick out?

- ☐ Does it make it muddy sounding?

G) What does it sound like if you set the LPF to 5kHz?

- ☐ Does it sit better in the mix?

H) What does the delay sound like on the lead vocal instead of the snare?

▶ Exercise 10.2: Setting A Delay From 50 To 100ms

A) Solo the Snare channel(s) and delay return again and and double the delay time used in the previous exercise until the delay time is between 50 and 100ms with no repeats.

☐ Does the Snare sound bigger?

☐ Does it sound more distinct?

☐ Can you hear the slap?

☐ Does it sound thicker?

B) Solo the rest of the drums.

☐ Do you have to increase the level to hear the delay?

☐ Does the Snare sound bigger in the mix?

☐ Can you hear a difference when you mute the delay?

☐ Does it sound more distinct?

C) Set the Feedback to about 20%.

☐ Does it sound better or worse?

☐ Can you hear the repeats

☐ Does it blend in with the mix or does it stick out?

☐ Does it make it muddy sounding?

D) What does it sound like if you set the LPF to 5kHz?

☐ Does it sit better in the mix?

E) What does the delay sound like on the lead vocal instead of the snare?

_____

▶ Exercise 10.3: Setting A Delay From 100 To 200ms

A) Solo the Snare channel(s) and delay returns again and double the delay time until it's between 100 and 200ms with no repeats.

☐ Does the snare sound bigger?

☐ Does it sound more distinct?

☐ Is it more distant?

☐ Is it distracting?

☐ Does it seem closer or further away from you?

B) Solo the rest of the drums.

☐ Do you have to increase the level to hear the delay?

☐ Can you hear a difference when you mute the delay?

☐ Does the Snare sound bigger in the mix?

☐ Does it sound more distinct?

☐ Is it more distant?

☐ Is it distracting?

C) Set the Feedback to about 20%.

☐ Does it sound better or worse?

☐ Does it make it muddy sounding?

D) What does it sound like if you set the LPF to 5kHz?

☐ Does it sit better in the mix?

E) What does the delay sound like with the lead vocal instead of the snare?

_____

> **Exercise 10.4:** _Setting A Delay Time Of 200 To 400ms_

A) Solo the Snare channel(s) and delay returns again and set the delay for whatever time denomination (it should be an ⅛th note) that sets the delay time between 200 and 400ms with no repeats.

☐ Does the snare sound bigger?

☐ Does it sound more distinct?

☐ Is it more distant?

☐ Is it distracting?

B) Solo the rest of the drums.

☐ Do you have to increase the level to hear the delay?

☐ Can you hear a difference when you mute the delay?

☐ Does the Snare sound bigger in the mix?

☐ Does it sound more distinct?

☐ Is it more distant?

☐ Is it distracting?

C) Set the Feedback to about 20%.

☐ Does it sound better or worse?

D) What does it sound like if you set the LPF to 5kHz?

☐ Does it sit better in the mix?

E) What does it sound like with the lead vocal instead of the snare?

_____

▶ *Exercise 10.5:* *Setting A Delay Time With A Triplet Note Value*

A) Solo the Snare channel(s) and delay returns again and set the delay for a *triplet* time denomination (⅛th note triplet) that sets the delay time between 200 and 400ms with no repeats.

☐ Does the snare sound bigger?

☐ Does it sound more distinct?

☐ Is it more distant?

☐ Is it distracting?

B) Solo the rest of the drums.

☐ Do you have to increase the level to hear the delay?

☐ Can you hear a difference when you mute the delay?

☐ Does the Snare sound bigger in the mix?

☐ Does it sound more distinct?

☐ Is it more distant?

☐ Is it distracting?

C) Set the Feedback to about 20%.

☐ Does it sound better or worse?

D) What does it sound like if you set the LPF to 5kHz?

☐ Does it sit better in the mix?

E) What does it sound like with the lead vocal instead of the snare?

▶ *Exercise 10.6:* *Setting A Delay With A Dotted Note Value*

A) Solo the Snare channel(s) and delay returns again and set the delay for a *dotted* time denomination (dotted 1/16th note) that sets the delay time between 200 and 400ms with no repeats.

☐ Does the snare sound bigger?

☐ Does it sound more distinct?

☐ Is it more distant?

☐ Is it distracting?

B) Solo the rest of the drums.

☐ Do you have to increase the level to hear the delay?

☐ Can you hear a difference when you mute the delay?

☐ Does the Snare sound bigger in the mix?

☐ Does it sound more distinct?

☐ Is it more distant?

☐ Is it distracting?

C) Set the Feedback to about 20%.
☐ Does it sound better or worse?

D) What does it sound like if you set the LPF to 5kHz?
☐ Does it sit better in the mix?

E) What does it sound like with the lead vocal instead of the snare?

Delay Setup

Just like with your reverbs, setting up the delays before you begin to mix can save a lot of time and distraction later. Here a setup to start you off, but eventually you'll develop your own preferences.

The Three Delay Full Setup Method

In this setup we'll use three different delays; one set up for a very short delay, another as a medium, and another long (see Figure 10.3). This covers most of the possibilities that might arise during a mix, although it's also possible that one dedicated to a specific mix element (like a solo or lead vocal) might be needed.

Figure 10.3: The Three Delay Setup

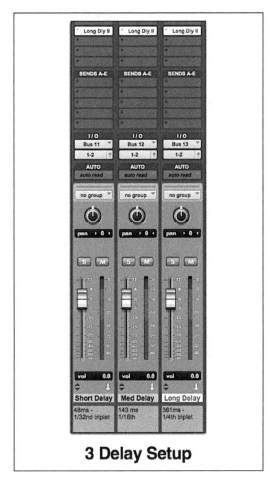

- **Delay One** - Set this up for a Haas Effect delay of less than 40ms and timed to the tempo of the song with no repeats. It will be used to make a track sound bigger and wider.

- **Delay Two** - Set this one up for a short delay of anywhere from 100 to 200ms with a couple of repeats and timed to the tempo of the song. This is used for a double or slap delay.

- **Delay Three** - Set this for for a long delay that will act as your "glue." Shoot for a timed delay that that goes anywhere from 250 to about 400ms with a couple of repeats. I like to use a triplet or dotted note delay here, but experiment to see what works for the song.

TIP: *If you want the delay to stick out, sometimes a 350ms untimed delay works great.*

These delay settings are just a starting place and will be tweaked as you go along in the mix. Sometimes you may find one of the delay times unnecessary, while other times you may need to add more.

Adding Delay To Other Mix Elements

Just like reverb, there's no rule for where and how delay is added to a mix. In some mixes, a single delay can work for every instrument and vocal (just like from all those great hits of 50s, 60s and 70s when the delay came from a tape machine), while another mix might sound better with a separate delay for many of the mix elements. Here are a few exercises that will allow you to hear each scenario.

Adding Delay To The Vocals

Both lead and background vocals are frequently the primary recipients of some sort of delay in the mix. For a lead vocal, it can provide a sense of space and polish without pushing the vocal too far back in the mix. For the background vocals, it can be a way to distinguish them from the lead vocal.

▶ *Exercise 10.7*: *Adding A Haas Delay To The Vocals*
Solo the lead Vocal and delay returns again.

A) Raise the level of the send to the Delay #1 with the Haas Effect setting.

☐ Does the vocal sound bigger?

☐ Is it more distant?

☐ Is it distracting?

☐ Is it thicker?

B) Unsolo the lead vocal and the delay returns.

☐ Does the lead vocal sound bigger in the mix?

☐ Is it more distant?

☐ Can you hear the delay?

☐ Is it distracting?

C) What does it sound like if you lower the delay level until you can just hear it?

D) What does it sound like if you add repeats?

E) What does it sound like if you set the LPF to 5kHz?

☐ Does it sit better in the mix?

F) What does it sound like if you pan the dry vocal to one side and delay to the other?

> **Exercise 10.8**: *Set A Slap Delay On A Vocal*

A) Raise the level of the send to the Delay #2 with the slap/double setting.

☐ Does the vocal sound bigger?

☐ Is it more distant?

☐ Is it distracting?

☐ Is it thicker?

B) Unsolo the lead Vocal and the delay returns.

☐ Does the lead vocal sound bigger in the mix?

☐ Is it more distant?

☐ Can you hear the delay?

☐ Is it distracting?

C) What does it sound like if you lower the delay level until you can just hear it?

D) What does it sound like if you add repeats?

E) What does it sound like if you set the LPF to 5kHz?

☐ Does it sit better in the mix?

F) What does it sound like if you pan the dry vocal to one side and delay to the other?

▶ *Exercise 10.9: Adding A Longer Delay To A Vocal*

A) Raise the level of the send to the Delay #3 with the longer delay setting.

☐ Does the vocal sound bigger?

☐ Is it more distant?

☐ Is it distracting?

☐ Is it thicker?

B) Unsolo the lead Vocal and the delay returns.

☐ Does the lead vocal sound bigger in the mix?

☐ Is it more distant?

☐ Can you hear the delay?

☐ Is it distracting?

C) What does it sound like if you lower the delay level until you can just hear it?

D) What does it sound like if you add repeats?

E) What does it sound like if you set the LPF to 5kHz?

☐ Does it sit better in the mix?

F) What does it sound like if you pan the dry vocal to one side and delay to the other?

Adding Delay To The Guitar

Like the vocal, both acoustic and electric guitars can benefit greatly by adding a bit of delay. It can make them larger than life, provide a sense of space, or push them back in the mix, depending on what the song requires and what you hear in your head.

▶ *Exercise 10.10: Adding Delay To The Guitar*

A) Repeat exercises E10.7, 10.8 and 10.9 using the guitar instead of the lead vocal.

▶ *Exercise 10.11: Adding A Room Delay To Guitar*

A) To set up a Guitar Room Delay:

• Insert a stereo delay into a guitar channel.

• Set the left delay for 25ms, the right for 50ms

• Set the Feedback to 10%

• Set the *Dry/Wet* control to 40%.

☐ Does the guitar sound bigger?

☐ Is it pushed back in the mix?

☐ Can you hear the delay?

☐ What happens when you adjust the *Feedback* or *Dry/Wet* parameters?

Adding Delay To Keyboards

All the attributes that a delay brings to a guitar can also be applied to just about any keyboard. While reverb may work better in some situations, never hesitate to experiment with a delay when looking for a sound. The result can be surprising.

▶ ***Exercise 10.12****: Adding Delay To keyboards*

A) Repeat exercises E8.7, 8.8 and 8.9 using the piano, organ or synthesizer instead of the lead vocal.

▶ ***Exercise 10.13****: Adding A Room Sound Delay To Keyboards*

A) To set up a Keyboard Room Delay:

• Insert a stereo delay into a keyboard channel.

• Set the left delay for 211ms, the right for 222ms

• Set the Feedback to 10%

• Set the *Dry/Wet* control to 40%. Does the keyboard sound bigger?

☐ Is it pushed back in the mix?

☐ Does it sound like it's in a room?

☐ What happens when you adjust the *Feedback* or *Dry/Wet* parameters?

☐ Can you hear the delays?

Adding Delay To The Drum Kit

While delay isn't used much as a primary effect on drums very much, it can be used for the glue effect when trying to tie everything in the mix sonically together. If a little bit of the delay that's used on one of the other instruments is added to the snare, it can provide a sense of movement since it can feel like the natural first reflection of a room. This works with toms and even the high-hat too, although not too well with cymbals.

▶ Exercise 10.14: Adding A Haas Delay To The Snare Drum

A) Solo the drum kit, then raise the send on the Snare channel(s) to Delay #1 with the Haas effect.

☐ Does the snare sound bigger?

☐ Does it sound smoother?

☐ Does it sound like it's in a different space?

☐ Does it fit better with the mix?

☐ Can you hear the difference when the send is muted?

☐ Can you hear it in the full mix?

B) Raise the send on the Tom channels to Delay #1.

☐ Do the toms sound bigger?

☐ Does the delay blend into the track?

☐ Does it sound like it's in a different space?

☐ Can you hear the difference when the send is muted?

☐ Does it fit better with the mix?

C) Raise the send on the Hat channel to Delay #1.

☐ Does the high-hat sound bigger?

☐ Does the delay blend into the track?

☐ Does it sound like it's in a different space?

☐ Can you hear the difference when the send is muted?

☐ Does it fit better with the mix?

▶ Exercise 10.15 Adding A Slap Delay To A Snare Drum

A) Solo the drum kit, then raise the send on the Snare channel(s) to Delay #2 with the slap/double effect.

☐ Does the snare sound bigger?

☐ Does it sound smoother?

☐ Does it sound like it's in a different space?

☐ Does it fit better with the mix?

☐ Can you hear the difference when the send is muted?

☐ Can you hear it in the full mix?

B) Raise the send on the Tom channels to Delay #1.

☐ Do the toms sound bigger?

☐ Does the delay blend into the track?

☐ Does it sound like it's in a different space?

☐ Can you hear the difference when the send is muted?

☐ Does it fit better with the mix?

C) Raise the send on the Hat channel to Delay #1 with the Haas Effect.

☐ Does the high-hat sound bigger?

☐ Does the delay blend into the track?

☐ Does it sound like it's in a different space?

☐ Can you hear the difference when the send is muted?

☐ Does it fit better with the mix?

> ### *Exercise 10.16: Adding Delay To The Drums*

A) Solo the drum kit, then raise the send on the Snare channel(s) to Delay #3 with the triplet "glue" effect.

☐ Does the snare sound bigger?

☐ Does it sound smoother?

☐ Does it sound like it's in a different space?

☐ Does it fit better with the mix?

☐ Can you hear the difference when the send is muted?

☐ Can you hear it in the full mix?

B) Raise the send on the Tom channels to Delay #1.

☐ Do the toms sound bigger?

☐ Does the delay blend into the track?

☐ Does it sound like it's in a different space?

☐ Can you hear the difference when the send is muted?

☐ Does it fit better with the mix?

C) Raise the send on the Hat channel to Delay #1 with the Haas Effect.

☐ Does the high-hat sound bigger?

☐ Does the delay blend into the track?

☐ Does it sound like it's in a different space?

☐ Can you hear the difference when the send is muted?

☐ Does it fit better with the mix?

Adding Delay To Other Mix Elements

While delay works great on many mix elements, there are a few it's rarely used on. You won't hear it much on bass, since it tends to rob the instrument of its power. It's not used much on percussion because it makes the many sharp attacks sound confusing (although you do hear it on congas every now and again). And you don't hear it much on strings, organ or any other pad element in the mix because the long sustaining notes negate the effect of any delay with repeats.

That being said, don't be afraid to try a delay on any instrument (especially a Haas Effect delay). You just might be surprised what it can do for the sound.

Chapter 11
Modulation Effects

Modulation refers to an external signal that varies the sound of an instrument or vocal in volume, timing or pitch. This includes effects like chorus, flanging and phasing, which are pretty standard mixing tools, to tremolo and vibrato, which are used mostly on guitars and electric pianos.

Between reverb, delay and modulation, modulation is the least used mixing effect because a little goes a long way. And except for flanging, you'll also hear modulation effects used mostly on a single track at a time, instead of across the entire mix. In fact, most modulation effects are inserted directly in the signal path of a channel instead of using a dedicated send and return configuration.

Modulation Basics

While all modulation effects certainly don't sound the same, not many mixers know the difference between them. Let's take a look.

Types Of Modulation

There are three modulation effects that are very closely related; phase shift, chorus, and flange. The simplest difference between them is that a chorus and flange effect comes from a modulated delay that's mixed back into the original signal. The flanger uses a shorter delay than a chorus, usually much less than 5 milliseconds, but a phaser uses no delay at all (see Figure 11.1).

Going a bit deeper, flangers, phasers and choruses work by producing a series of frequency bandwidth notches that are slowly swept across the frequency spectrum of the instrument or vocal. You don't really hear these notches; you hear what's left in the frequency spectrum, which is a series of peaks. Phasers have a small number of notches spaced evenly across the frequency spectrum while flangers and choruses have a larger number that are spaced harmonically.

Tremolo and vibrato work a little differently because no delay is involved. Tremolo cyclically varies the signal up and down in level, while vibrato varies the tone cyclically up and down.

Effect	Delay	Description
Phase Shift	None	Cancels different frequencies to create the effect. Frequency notches spaced evenly across the guitar frequency response.
Flanging	Less than 1ms to 5ms	Deepest depth have the greatest frequency cancellations. Frequency notches are spaced harmonically across the instrument's frequency response.
Chorus	5 to 25ms	Almost sounds like doubling while widening the sound. Used to thicken the sound and create a stereo image. Frequency notches are spaced harmonically across the instrument's frequency response.
Tremelo	None	Cyclicly changes the volume. Used mostly on guitars.
Vibrato	None	Cyclicly changes the frequency response.

Flangers And Phasers

The flanger is a dramatic effect that was first derived in the 60s by playback of a song on two tape recorders at the same time, and slowing one down by placing your finger on the tape flange, hence the name "flanging." One of the first hit songs that it was used on was a 1967 hit by the British group The Small Faces called "Itchycoo Park." which featured a large dose of the effect on each chorus. As these things usually go, once the song became a hit, every artist and producer wanted the effect on their song after that. The problem was that setting up the two tape recorders needed for the effect was both expensive and very time consuming, so it wasn't long before an electronic simulation of sorts soon came on the market.

Back in the 70s before the introduction of inexpensive digital delay electronics, an analog phaser was the only way to get any sort of modulated effect, but it was a weak cousin compared to the intensity of a deep flange, which is why phasing isn't used much - it's just not that dramatic an effect.

Once digital delays came on the market it became possible to simulate true tape flanging, and now just about every modulation plugin and stomp box does a great job.

Chorus

Chorus differs from flanging in a couple of aspects; the delay needed to achieve the effect is longer, and the frequency notches aren't random like in flanging. You can thank Roland and their Jazz Chorus line of guitar amplifiers introduced in 1980 for starting the chorus craze, which was soon included across their line of keyboards as the world became hooked on the effect. Soon after the amp received such a great response, Roland introduced a version in their Boss line of stomp boxes that engineers used to easily add the effect wherever needed during a mix (see Figure 11.2).

Figure 11.2: Boss CE-1 Chorus
Courtesy of Boss Corporation

Indeed, it's very easy to fall in love with the sound of a chorus because it's lush sounding, and if used in stereo, can really widen the sound of a track quite a bit. If you listen to many hits from the 80s, you'll hear the effect used, used, and used some more.

Today, most modulation plugins allow you to select between chorus, phasing and flanging, since they're all related.

Tremolo And Vibrato

A tremolo is a cyclic variation in volume, like what you find on Fender amplifiers (even though it's incorrectly labeled as "Vibrato"). Don't confuse tremolo with vibrato because they're different. Vibrato changes the pitch of the sound while tremolo changes the volume.

While tremolo is occasionally used on a guitar or electric piano (both the original Rhodes and Wurlitzer electronic pianos had it built in), vibrato is rarely used since the variation in pitch can make band tuning an issue.

Parameter Settings

The parameters for a phasor, flanger, or chorus, are somewhat the same, which is why all three effects are sometimes combined into the same plugin (see Figure 11.3).

Figure 11.3: Typical Modulation Parameter Controls
Courtesy of Universal Audio

Speed/Rate - This control adjusts the speed of the effect. Usually it's set in the low milliseconds from 0.5 to about 25 cycles per second.

Depth/Intensity - Adjusts how much you hear the effect.

Mix - Sometimes a *Mix* control, which mixes the wet with the dry signal, is added instead of a *Depth* or *Intensity* control. In most cases, a mix of 50/50 provides the most intense effect.

Feedback/Regeneration - *Feedback* (sometimes called *Regeneration*) takes a little of the output signal of the plugin and routes it back into the input of the device, which provides an unusual variation to the sound.

Width - This controls how wide the stereo field is.

Delay - On many multi-function modulation devices or plugins, the *Delay* control is what changes the function from a phasor (no delay) to a flanger (0.5 to 5ms) to a chorus (25ms) or anywhere in between.

A variety of other controls, from input and output to equalizers and filters, can also be found on some of the more sophisticated devices.

Tremolo/Vibrato

Tremolo and vibrato plug-ins usually have two parameter settings; speed and depth.

Speed - In a tremolo, this control adjusts how fast the volume level will change. In a vibrato, it controls how fast the tuning will change.

Depth - This adjusts the how much of the effect you hear. Sometimes the depth is fixed so this control isn't included.

Modulation Setup

In the 80s when chorus seemed to be used on nearly mix element, you frequently found a chorus unit patched into its own console returns so it could be easily accessed by a send from any channel. Today we use a chorus only occasionally, and even less for the other modulation effects, so it's usually patched directly into the signal path of the channel that we want to effect, and the amount used is controlled by the Mix or Depth control.

On rare occasions, a song may require the entire mix to be flanged at some point, which would require it to be patched into the signal path of the stereo mix buss.

Modulation On Instruments

Unlike reverb and delay, modulated effects are used consistently on a few instruments more than others. Let's look at those first.

Modulation On Guitars

Distorted electric guitars and modulation seem to fit together like a hand in a glove, although clean electrics and acoustics benefit significantly as well. Want to widen it out in the stereo field? Add a little chorus. What it to sound thicker? Some chorus or a slight flange can help. Want that Eddie Van Halen sound from "Unchained?" That's a flanger he's using.

Tremolo is another go-to effect for guitar that's been used on many hit songs over the years like "Gimme' Shelter" by the Rolling Stones, "Born On The Bayou" by Creedence Clearwater Revival, and something a little more current, "The Bends" by Radiohead.

▶ *Exercise 11.1*: *Using Chorus On Guitars*

All exercises in the chapter will use the example song "It's About Time," but they could also apply to any other song that you want to work on.

Insert a chorus plugin into the signal path of a guitar channel. Set the Mix parameter to around 40%.

A) Solo the guitar channel and raise the *Rate* control.

☐ Does the guitar sound smoother or muddy with the chorus?

☐ Can you hear the difference when you bypass the chorus plugin?

☐ Does the chorus sound better at a slower or faster rate?

B) Raise the *Depth* control.

☐ Can you hear the difference when you bypass the chorus plugin?

C) Unsolo the guitar channel.

☐ Does the guitar fit into the mix better?

☐ Do you need to change the level of the chorus to hear it better?

☐ Does the sound of the guitar change too much for the song?

☐ Are there now tuning problems with any other mix element?

▶ *Exercise 11.2*: Using Flanging On Guitars

Replace the chorus plugin with a flanger plugin. Set the Mix parameter to around 40%.

A) Solo the guitar channel and raise the *Rate* control.

☐ Does the guitar sound smoother or muddy with the flanger?

☐ Can you hear the difference when you bypass the flanger plugin?

☐ Does the flanger sound better at a slower or faster rate?

B) Raise the *Depth* control.

☐ Can you hear the difference when you bypass the flanger plugin?

C) Raise the *Feedback* control.

☐ Does the sound of the guitar change too much for the song?

☐ Are there tuning problems with any other mix element?

☐ Does the flanger sound better when the feedback is increased?

D) Raise the *Mix* control.

☐ Does the sound of the effect change?

☐ Does the sound of the guitar get smaller or bigger?

B) Raise the *Depth* control.

☐ Can you hear the difference when you bypass the flanger plugin?

☐ Does the sound of the guitar get smaller or bigger?

C) Unsolo the guitar channel.

☐ Does the guitar fit into the mix better?

☐ Do you need to change the level of the flanger to hear it better?

☐ Does the sound of the guitar change too much for the song?

☐ Are there tuning problems with any other mix element?

▶ *Exercise 11.3*: Using Tremolo On Guitars

Replace the flanger plugin with a tremolo plugin (if your DAW has one available).

A) Solo the guitar channel and raise the *Rate* control.

☐ Does the effect sound better timed to the tempo of the song?

☐ Can you hear the difference when you bypass the tremolo plugin?

☐ Does the tremolo sound better at a slower or faster rate?

B) Raise the *Depth* control.

☐ Can you hear the difference when you bypass the tremolo plugin?

C) Unsolo the guitar channel.

☐ Does the guitar fit into the mix better?

☐ Do you need to change the level of the tremolo to hear it better?

☐ Does the sound of the guitar change too much for the song?

☐ Are there now tuning problems with any other mix element?

Modulation On Keyboards

Chorus or flanging has been used on keyboards almost from the time they were invented. Either process can take a boring mono instrument and make it interesting, as well as make it thicker and wider sounding. Be careful not to use too much though (especially on a grand piano), since it not only can change the character of the sound to something that sounds dated, but it can cause a tuning problem with other instruments as well.

Tremolo goes very well with electric piano, especially one that cross-pans from the left channel to the right (or vice-versa). The very earliest electric pianos had tremolo built-in, and if you try it a little, you'll know why.

▶ *Exercise 11.4*: Using Chorus On Keyboards

Insert a chorus plugin into the signal path a keyboard channel of your DAW. Set the Mix parameter to around 40%.

A) Solo the keyboard channel and raise the *Rate* control.

☐ Does the keyboard sound smoother or muddy with the chorus?

☐ Can you hear the difference when you bypass the chorus plugin?

☐ Does the chorus sound better at a slower or faster rate?

B) Raise the *Depth* control.

☐ Can you hear the difference when you bypass the chorus plugin?

C) Unsolo the guitar channel.

☐ Does the keyboard fit into the mix better?

☐ Do you need to change the level of the chorus to hear it better?

☐ Does the sound of the keyboard change too much for the song?

☐ Are there now tuning problems with any other mix element?

▶ **Exercise 11.5**: *Using A Flanger On Keyboards*

Replace the chorus plugin with a flanger plugin. Set the Mix parameter to around 40%.

A) Solo the keyboard channel and raise the *Rate* control.

☐ Does the keyboard sound smoother or muddy with the flanger?

☐ Can you hear the difference when you bypass the flanger plugin?

☐ Does the flanger sound better at a slower or faster rate?

B) Raise the *Depth* control.

☐ Can you hear the difference when you bypass the flanger plugin?

C) Raise the *Feedback* control.

☐ Does the sound of the keyboard change too much for the song?

☐ Are there tuning problems with any other mix element?

☐ Does the flanger sound better when the feedback is increased?

D) Raise the *Mix* control.

☐ Does the sound of the effect change?

☐ Does the sound of the guitar get smaller or bigger?

B) Raise the *Depth* control.

☐ Can you hear the difference when you bypass the flanger plugin?

☐ Does the sound of the keyboard get smaller or bigger?

C) Unsolo the keyboard channel.

☐ Does the keyboard fit into the mix better?

☐ Do you need to change the level of the flanger to hear it better?

☐ Does the sound of the keyboard change too much for the song?

☐ Are there tuning problems with any other mix element?

▶ Exercise 11.6: Using A Tremelo On Keyboards

Replace the flanger plugin with a tremolo plugin (if your DAW has one available).

A) Solo the keyboard channel and raise the *Rate* control.

☐ Does the effect sound better timed to the tempo of the song?

☐ Can you hear the difference when you bypass the tremolo plugin?

☐ Does the tremolo sound better at a slower or faster rate?

B) Raise the *Depth* control.

☐ Can you hear the difference when you bypass the tremolo plugin?

C) Unsolo the keyboard channel.

☐ Does the keyboard fit into the mix better?

☐ Do you need to change the level of the tremolo to hear it better?

☐ Does the sound of the keyboard change too much for the song?

☐ Are there now tuning problems with any other mix element?

Modulation On Vocals

Sometimes chorus is used on a lead vocal, but in a very subtle way. A chorus is brought back into the console/DAW on two separate channels besides the vocal, and then spread out slightly across the stereo spectrum with the level raised so it's just behind the original vocal, Not only will the lead vocal sound bigger, but any tuning inconsistencies can be covered up as well.

TIP: Chorus or a slight flange on background vocals can also make them sounder thicker and separate them from the lead vocal.

▶ Exercise 11.7: Using A Chorus On Lead Vocal

Insert a stereo chorus plugin into an aux channel on the DAW and assign an aux send to it. Pan the return all the way to the left and right and set the mix to 100%.

A) Solo the lead Vocal channel and raise the level of the send to the chorus.

☐ Can you hear the stereo effect?

☐ Can you hear the difference when you mute the chorus channel?

B) Pan the Chorus aux channel to the 11 and 1 o'clock positions.

☐ Does the vocal sound more or less focused?

C) Unsolo the Vocal channel and raise the *Rate* control.
☐ Does the vocal fit into the mix better?

☐ Do you have to change the level of the chorus to hear it?

☐ Does it sound better if the chorus is timed to the tempo of the song?

☐ Does it sound better if the rate is slower?

D) Raise the *Depth* control.
☐ Does the vocal fit into the mix better?

☐ Does the sound of the vocal change too much for the song?

☐ Does it sound better with less chorus?

☐ Are there tuning problems with any other mix element?

➤ Exercise 11.8: Using A Flanger On Lead Vocal

Replace the chorus plugin with a flanger plugin. Pan the returns all the way to the left and right and set the mix to 100%.

A) Solo the lead Vocal channel and raise the level of the send to the flanger.
☐ Can you hear the stereo effect?

☐ Does the vocal sound thicker?

☐ Does it sound better if you lower the flanger level?

B) Pan the flanger aux channel to the 11 and 1 o'clock positions.
☐ Does the vocal sound more or less focused?

C) Unsolo the Vocal channel and raise the *Rate* control.
☐ Does the vocal fit into the mix better?

☐ Do you have to change the level of the flanger to hear it?

☐ Does it sound better if the flanger is timed to the tempo of the song?

☐ Does it sound better if the rate is slower?

D) Raise the *Depth* control.
☐ Does the vocal fit into the mix better?

☐ Does the sound of the vocal change too much for the song?

☐ Does it sound better with less flange?

☐ Are there tuning problems with any other mix element?

E) Raise the *Feedback* control.
☐ Does the sound fit into the mix better?

☐ Does the sound of the lead vocal change too much for the song?

☐ Is there tuning problems with any other mix element?

☐ Does it sound better or worse if the feedback is increased.

▶ Exercise 11.9: Using A Flanger On Background Vocals

Insert an aux send on the background vocals subgroup or channels and send it to the flanger.

A) Solo the background vocal channels and raise the level of the send to the flanger.

☐ Can you hear the stereo effect?

☐ Is the stereo spread of the background vocals more pronounced?

☐ Do the background vocals sound thicker?

☐ Does it sound better if you lower the flanger level?

B) Pan the flanger aux channel to the 11 and 1 o'clock positions.

☐ Do the background vocals sound more or less focused?

C) Unsolo the background vocal channel and raise the *Rate* control.

☐ Do the background vocals fit into the mix better?

☐ Do you have to change the level of the flanger to hear it?

☐ Does it sound better if the flanger is timed to the tempo of the song?

☐ Does it sound better if the rate is slower?

D) Raise the *Depth* control.

☐ Do the background vocals fit into the mix better?

☐ Does the sound of the background vocals change too much for the song?

☐ Does it sound better with less flange?

☐ Are there tuning problems with any other mix element?

E) Raise the *Feedback* control.

☐ Does the sound fit into the mix better?

☐ Does the sound of the background vocals change too much for the song?

☐ Is there tuning problems with any other mix element?

☐ Does it sound better or worse if the feedback is increased.

_____

▶ Exercise 11.10: Using Chorus On Background Vocals

Replace the flanger plugin with a chorus plugin.

A) Solo the background vocal channels and raise the level of the send to the chorus.

☐ Can you hear the stereo effect?

☐ Can you hear the difference when you mute the chorus channel?

B) Pan the Chorus aux channel to the 11 and 1 o'clock positions.

☐ Do the background vocals more or less focused?

C) Unsolo the Vocal channel and raise the *Rate* control.

☐ Do the background vocals fit into the mix better?

☐ Do you have to change the level of the chorus to hear it?

☐ Does it sound better if the chorus is timed to the tempo of the song?

☐ Does it sound better if the rate is slower?

D) Raise the *Depth* control.

☐ Do the background vocals fit into the mix better?

☐ Does the sound of the background vocals change too much for the song?

☐ Does it sound better with less chorus?

☐ Are there tuning problems with any other mix element?

Modulation On Strings

Strings and modulation go together like peanut butter and jelly. Sometimes a little chorus on a small string section can make it sound a lot larger than it is, and flanging on strings makes for a very dramatic effect.

▶ *Exercise 11.11*: *Using A Chorus On Strings*

Insert a chorus plugin into the string channels or subgroup of your DAW, or send to the aux track that contains the chorus plugin.

A) Raise the Rate control.

☐ Do the strings fit into the mix better?

☐ Do the strings sound thicker?

☐ Can you hear the difference when you mute the chorus channel?

☐ Does it sound better if the rate is slower?

B) Raise the Depth control.

☐ Do the strings fit into the mix better?

☐ Does the sound of the strings change too much for the song?

☐ Are there tuning problems with any other mix element?

▶ Exercise 11.12: Using A Flanger On Strings

Insert a flanger plugin into the string channels or subgroup of your DAW, or send to the aux track that contains the flanger plugin.

A) Raise the Rate control.
- ☐ Do the strings fit into the mix better?
- ☐ Do the strings sound thicker?
- ☐ Can you hear the difference when you mute the flanger channel?
- ☐ Does it sound better if the rate is slower?

B) Raise the Depth control.
- ☐ Do the strings fit into the mix better?
- ☐ Does the sound of the strings change too much for the song?
- ☐ Are there tuning problems with any other instrument?

C) Raise the Feedback control.
- ☐ Do the strings fit into the mix better?
- ☐ Does the sound of the strings change too much for the song?
- ☐ Are there tuning problems with any other mix element?
- ☐ Does it sound better if the feedback is increased?

Modulation On Other Mix Elements

There are some mix elements that don't lend themselves to modulation. Drums are a good example. Because of the short bursts of energy from a drum hit, most modulation effects aren't that apparent, with the exception of the cymbals.

Sometimes a bit of chorus can be used to thicken up a snare a bit, as well as the bass, as we'll hear shortly. A little goes a long way though, as modulation changes the tone of the bass and as a result, can take away some of its power, which is so important to the song.

▶ Exercise 11.13: Using A Chorus On The Snare Drum

Insert a chorus plugin into the snare channels or subgroup of your DAW, or send to the aux track that contains the chorus plugin.

A) Solo the Snare channel and raise the level of the send to the chorus.
- ☐ Can you hear the stereo effect?
- ☐ Can you hear the difference when you mute the chorus channel?

B) Pan the Chorus aux channel to the 11 and 1 o'clock positions.
- ☐ Does the snare sound more or less focused?

C) Unsolo the Snare channel and raise the *Rate* control.

☐ Does the snare fit into the mix better?

☐ Do you have to change the level of the chorus to hear it?

☐ Does it sound better if the rate is slower?

D) Raise the *Depth* control.

☐ Does the snare fit into the mix better?

☐ Does the sound of the snare change too much for the song?

☐ Does it sound better with less chorus?

▶ **Exercise 11.14**: *Using A Flanger On Snare Drum*

Insert a flanger plugin into the snare channels or subgroup of your DAW, or send to the aux track that contains the flanger plugin.

A) Solo the Snare channel and raise the level of the send to the flanger.

☐ Can you hear the stereo effect?

☐ Does the snare sound thicker?

☐ Does it sound better if you lower the flanger level?

B) Pan the flanger aux channel to the 11 and 1 o'clock positions.

☐ Does the snare sound more or less focused?

C) Unsolo the Snare channel and raise the *Rate* control.

☐ Does the snare fit into the mix better?

☐ Do you have to change the level of the flanger to hear it?

☐ Does it sound better if the rate is slower?

D) Raise the *Depth* control.

☐ Does the snare fit into the mix better?

☐ Does the sound of the vocal change too much for the song?

☐ Does it sound better with less flange?

E) Raise the *Feedback* control.

☐ Does the sound fit into the mix better?

☐ Does the sound of the snare change too much for the song?

☐ Does it sound better or worse if the feedback is increased.

▶ *Exercise 11.15*: Using A Chorus On Cymbals

Insert a chorus plugin into the cymbal channels or subgroup of your DAW, or send to the aux track that contains the chorus plugin.

A) Solo the cymbal channels and raise the level of the send to the chorus.

☐ Can you hear the stereo effect?

☐ Can you hear the difference when you mute the chorus channel?

B) Pan the Chorus aux channel to the 11 and 1 o'clock positions.

☐ Do the cymbals sound more or less focused?

C) Unsolo the Vocal channel and raise the *Rate* control.

☐ Do the cymbals fit into the mix better?

☐ Do you have to change the level of the chorus to hear it?

☐ Does it sound better if the rate is slower?

D) Raise the *Depth* control.

☐ Do the cymbals fit into the mix better?

☐ Does the sound of the cymbals change too much for the song?

☐ Does it sound better with less chorus?

▶ *Exercise 11.16*: Using A Flanger On Cymbals

Insert a flanger plugin into the cymbal channels or subgroup of your DAW, or send to the aux track that contains the flanger plugin.

A) Solo the cymbal channels and raise the level of the send to the flanger.

☐ Can you hear the stereo effect?

☐ Do the cymbals sound thicker?

☐ Does it sound better if you lower the flanger level?

B) Pan the flanger aux channel to the 11 and 1 o'clock positions.

☐ Do the cymbals sound more or less focused?

C) Unsolo the Vocal channel and raise the *Rate* control.

☐ Do the cymbals fit into the mix better?

☐ Do you have to change the level of the flanger to hear it?

☐ Does it sound better if the rate is slower?

D) Raise the *Depth* control.

☐ Do the cymbals fit into the mix better?

☐ Does the sound of the the cymbals change too much for the song?

☐ Does it sound better with less flange?

E) Raise the *Feedback* control.

☐ Does the sound fit into the mix better?

☐ Does the sound of the lead vocal change too much for the song?

☐ Does it sound better or worse if the feedback is increased.

► Exercise 11.17: Using A Chorus On Bass

Insert a chorus plugin into the bass channels or subgroup of your DAW, or send to the aux track that contains the chorus plugin.

A) Solo the bass channel and raise the level of the send to the chorus.

☐ Can you hear the stereo effect?

☐ Can you hear the difference when you mute the chorus channel?

B) Pan the Chorus aux channel to the 11 and 1 o'clock positions.

☐ Does the bass sound more or less focused?

C) Unsolo the Vocal channel and raise the *Rate* control.

☐ Does the bass fit into the mix better?

☐ Do you have to change the level of the chorus to hear it?

☐ Does it sound better if the rate is slower?

D) Raise the *Depth* control.

☐ Does the bass fit into the mix better?

☐ Does the sound of the vocal change too much for the song?

☐ Does it sound better with less chorus?

☐ Are there tuning problems with any other mix element?

In this chapter you've heard the difference between a chorus and flanger on various instruments. While the flanger is more dramatic, sometimes the chorus is a better choice for adding movement and thickening a sound. It's your choice, as there are no wrong answers here.

Chapter 12
Creating Interest

There's more to mixing than just balancing the instruments and vocals and adding some EQ and effects. To really make a mix rock, it has to both feel good and be interesting as well. How does this happen? By finding the tracks that establish the groove or demand your attention, then emphasizing them.

Developing The Groove

As stated in Chapter 3, the groove is the pulse of the song. While it usually comes from the drums and bass, it could really come from any instrument or even a vocal. And lest you think that the groove is predominantly a fixture of one type of music like funk or R&B, you'll find that a strong groove exists in just about any type of good music, regardless of the genre or style. Don't believe me? Listen to the US Marine Corps band play "Stars And Stripes Forever" and then listen to a typical high school band. The Marines have a groove that makes you want to jump up and march with them just as much as you want to shake your booty to a James Brown or Prince song.

Finding The Groove

The best way to develop the groove is to find the instrument or instruments in the song that supplies its pulse. As said before, it's usually the bass and drums, but it could very well be a loop, a keyboard, a guitar, and rarely, a vocal. If a band is playing particularly well together, it may be several instruments at the same time.

▶ ***Exercise 12.1:*** *Finding The Groove*

A) Refer to the mix that you've been working on in the previous chapters.

☐ Listen to the entire mix all the way through. Is there an instrument or instruments that establishes the pulse of the song?

☐ If an instrument doesn't stick out as being the pulse of the song, go through each track one by one and raise the level by about 3dB. After you've listened, return it to its previous level position. Do you hear one or more tracks as establishing the pulse of the song now?

Understand that if you're working on a song that wasn't well performed, there may not be a track that establishes the groove. If that's the case, it's usually the producer's call to recut the track, or you'll just have to make due with the instrument that feels the best.

TIP: Remember that mixing is always easier with well recorded tracks, great playing, and excellent arrangements.

Establishing The Groove

Once the instrument (or instruments) that establishes the groove is found, the next step is to emphasis it. This can be done by raising the level as little as one dB, or adding an extra bit of compression or EQ to make it stand out a bit more in the mix. Then make sure that the rest of the tracks support your groove mix element by tailoring the mix around it.

Take notice in the exercises below that we're adding very small increments of level, compression or EQ. In theory, 1dB is the minimum amount that the average person can hear according to most text books (it's actually less than that), and sometimes that's all you need to change the balance or feel in a mix. On just about anything in mixing, always begin with small increments first.

▶ *Exercise 12.2: Establishing The Groove*

After your groove tracks are found, try one or more of the following:

A) Raise the level of each of the groove tracks by 1dB.

☐ Can you feel the groove better?

B) If not, add another dB.

☐ Can you feel it now?

C) If not, add another dB, but be cautious not to make the tracks too loud in the mix.

D) If your groove tracks are already being compressed, add another dB or two of compression. Keep the level the same by adjusting the *Output* control or raising the channel fader.

☐ Can you feel the groove better? If the groove tracks aren't compressed, then refer back to Chapter 7 and add compression.

E) If your groove tracks are already equalized, add an additional dB at the EQ points.

☐ Can you feel the groove better? If not, add another dB.

☐ Can you feel it now? Be cautious that the tracks are not too far out in front of the mix or that they don't clash with another instrument.

F) After the groove is established and drives the mix, do any final tweaks to the other tracks to make sure they're not covered up or that they don't clash with the groove tracks.

Emphasizing The Most Important Element

In every song one mix element is more important than everything else. In dance or electronic music it may be the groove but in a genre like traditional country music it's the vocal. Sometimes it's a riff (like Coldplay's "Clocks") and sometimes it's a loop (like gang vocal loop that Kenye West uses in "Power"). The job of the mixer is to identify this element and emphasize it in the mix.

Finding The Most Important Element

The most important element of the song is the one that captures your ear and makes you think "Cool!" It could be a modulation or delay effect on an instrument or voice, it could be a unique instrument sound, it could be an interesting hook or riff, or it could be an especially passionate vocal performance. The whole trick is to find that part, then emphasize it.

▶ *Exercise 12.3: Finding The Most Important Element*

Refer to the mix that you've been working on in the previous chapters.

A) Listen to the entire mix all the way through.

☐ Is there an instrument or instruments that captures your ear as unique or interesting?

B) Is there one that's *potentially* interesting?

☐ Would adding an effect grab your attention?

C) If an instrument doesn't stick out as being the most interesting in the song, go through each track one by one and raise it by about 3dB.

D) After you've listened, return it to its previous level position.

☐ Do you hear a track that's interesting or potentially interesting now?

_____

Emphasizing The Most Important Element

Once you've found your interesting element, now it's time to make sure that it pulls the listener into the song immediately. Like with the groove, this can be done with either level, compression, EQ or effects.

▶ **Exercise 12.4:** *Emphasizing The Most Important Element*

After your interesting element is found, try one or more of the following:

A) Raise the level of the mix element by 1dB.

☐ Does it jump out of the mix more?

☐ Does it grab the listener's attention?

B) If not, add another dB.

☐ Does it grab your attention now?

C) If not, add another dB, but be cautious that the track isn't too far out in front of the mix.

D) If the selected track is already being compressed, add another dB or two of compression. Keep the level the same by readjusting the *Output* control.

☐ Does it stand out more in the mix?

☐ Does it grab the listener's attention now?

E) If the selected track isn't compressed, then refer back to Chapter 7 and add compression.

F) If your selected track is already equalized, add an additional dB at the EQ points.

☐ Does it grab the listener's attention?

G) If not, add another dB.

☐ Does it jump out of the mix now? Be cautious that the tracks are not too far out in front of the mix or that they don't clash with another instrument.

H) Add 1dB at 5kHz.

☐ Does it grab the listener's attention?

I) If not, add another dB.

☐ Does it jump out of the mix now?

Making A Mix Element Interesting

In the case of finding a potentially interesting element, you'll have to work a little harder, since it's up to you to take it over the top and make it interesting. This usually takes a lot of experimenting and can take up the bulk of the time during a mix.

Many times radical EQ or rarely used effects like ping-pong delays or a flange with massive regeneration can be just the thing to bring a track to life. You'll never know until you try it with the track.

▶ **Exercise 12.5:** *Making A Mix Element Interesting With EQ*

After you've selected a mix element, try radically EQing the track.

A) Add an LPF to the channel if there's not one already inserted. Gradually lower the frequency.

☐ Is there a point where the track becomes more interesting?

B) Add a HPF to the channel if there's not one already inserted. Gradually increase the frequency.

☐ Is there a point where the track becomes more interesting?

C) Sweep the midrange of the track with an EQ boosted by 6dB.

☐ Is there a frequency that jumps out?

D) What happens if you boost that frequency even more?

E) What happens if you cut it?

☐ Is it more interesting?

F) Add a few dB at 5kHz.

☐ Does it grab the listener's attention?

G) Add a few dB at 12kHz.

☐ Does it grab the listener's attention?

H) Try a different style of EQ (such as a tube emulation) and repeat steps A through G.

_____

▶ **Exercise 12.6:** *Making A Mix Element Interesting With Compression*

After you've selected an element, try radically compressing the track.

A) Increase the *Threshold* control to where it peaks at about 10dB of compression.

B) Keep the level the same by readjusting the *Output* control.

    ☐ Does it catch the listener's ear?

    ☐ How about 20dB?

C) Set the *Ratio* control to 10:1. Keep the level the same by readjusting the *Output* control.

    ☐ Does it catch the listener's ear?

D) Now set the ratio to 20:1.

    ☐ Does it catch the listener's ear?

E) Turn the *Attack* parameter to as fast as it will go.

    ☐ Does it sound bad or unique?

F) Turn it as slow as it will go.

    ☐ Does it sound bad or unique?

G) Turn the *Release* parameter to as fast as it will go.

    ☐ Does it sound bad or unique?

H) Turn it as slow as it will go.

    ☐ Does it sound bad or unique?

I) Try a different style of compressor and repeat steps A through H.

▶ **Exercise 12.7:** *Making A Mix Element Interesting With Delay*
After you've selected an element, try radically delaying the track. Insert a dedicated delay plug-in into the channel. Adjust the *Mix* control so that you can hear the delay without it being too loud.

A) Time the delay for a quarter note delay as described in Chapter 10.
    ☐ Is the track now more or less interesting?

    ☐ Does it sound more interesting with more or fewer repeats?

B) Time the delay for an eighth note delay.
    ☐ Is the track now more or less interesting?

    ☐ Does it sound more interesting with more or fewer repeats?

C) Try other timed note denominations like 1/16th, 1/32nd, 1/64th, etc.
    ☐ Try more or fewer repeats.

    ☐ Is the track now more or less interesting?

D) Now radically filtered out the highs and lows of the effect.

☐ Is the track now more or less interesting?

E) Try different triplet note denominations.

☐ Is the track now more or less interesting?

F) different dotted note denominations.

☐ Is the track now more or less interesting?

G) Try a delay that's set to 350ms and not timed to the track.

☐ Is the track more or less interesting?

☐ 150ms?

☐ 75ms?

H) Try a stereo delay and pan the two delays hard left and right. Set the delay times differently on each side (for example an 1/8th note on the left and a 1/16th note on the right).

☐ Is the track now more or less interesting?

I) Try a ping-pong delay or any other preset that the delay plug-in may have. Tailor it to the track.

☐ Is the track now more or less interesting?

▶ *Exercise 12.8: Making A Mix Element Interesting With Reverb*

After you've selected a mix element, try adding some radical reverb to the track. Insert a dedicated reverb into the signal path of the the channel. Adjust the *Mix* control so that you can hear the reverb without it being too loud.

A) Set the reverb time as short as it will go.

☐ Is the track more interesting?

B) Set the reverb time as long as it will go.

☐ Is the track more interesting?

C) Slowly decrease the LPF frequency on the reverb if there is one.

☐ Is there a point where it becomes more interesting?

D) Slowly increase the HPF frequency on the reverb if there is one.

☐ Is there a point where it becomes more interesting?

E) Increase the pre-delay from 0ms to as long as it will go.

☐ Is there a point where it becomes more interesting?

F) Try different types of reverbs (hall, chamber, room, plate) and repeat steps A through E.

☐ Is there a point where it becomes more interesting?

> ### Exercise 12.9: *Making A Mix Element Interesting With A Modulation Effect*

After you've selected a mix element, insert a dedicated modulator, like a chorus or flanger, into the signal path of the channel.

A) Adjust the *Mix* control so the effect is very soft to radically loud.

☐ Is there a point where it becomes more interesting?

B) Adjust the *Rate* control so the effect is very slow to radically fast.

☐ Is there a point where it becomes more interesting?

C) Adjust the *Depth* control so the effect is very soft to radically loud.

☐ Is there a point where it becomes more interesting?

D) Adjust the *Feedback* control so the effect goes from 0 to full level.

☐ Is there a point where it becomes more interesting?

E) Adjust the *Width* control so the effect goes from wide to narrow.

☐ Is there a point where it becomes more interesting?

F) Change the setting to flange (or insert a flanger plug-in), and repeat steps A through E.

☐ Does changing to another modulator make it more interesting?

Using Saturation

Sometimes saturation can be the secret sauce in making a mix element more interesting, or simply fitting into a mix better. Saturation actually comes in two kinds of flavors - tape saturation and overdrive saturation. Tape saturation is used to add the glue to a mix that analog tape might have done back in the day, but that's not going to make the sound more interesting. Overdrive saturation is what we're looking for here.

Overdrive saturation plugins can go from the very simple to the very complex. Most have two main parameters though, *Overdrive* (or *Saturation*), and *Tone Color*. There can be a variety of other parameters mixed in, but most are just adjustment to the main two.

The best thing about overdrive saturation is that it can make a mix element absolutely jump out of a mix in a way that can't be done with compression or EQ. Let's look at how to do that.

▶ *Exercise 12.10: Making A Mix Element Interesting With Saturation*

After you've selected a mix element, insert a dedicated saturation plugin into the channel. Adjust the *Mix* control to about 50% wet.

A) Raise the *Saturation* or *Overdrive* control until you begin to hear the sound of the mix element change.

☐ Can the mix element be heard better in the mix?

☐ Did the sound of the mix element change?

B) Raise the *Saturation* or *Overdrive* control until it sounds fuzzy or distorted.

☐ Can the mix element be heard better in the mix?

☐ Does the mix element sound more interesting or worse than before?

C) Reduce the *Saturation* or *Overdrive* control until you can just barely hear it on the mix element. Go to the *Color* or *Tone* control (if it has one) and increase it half-way.

☐ Can the mix element be heard better in the mix?

☐ Does the mix element sound more interesting or worse than before?

D) If there are additional tone controls, continue experimenting to see if the mix element fits in the mix better or sounds more interesting.

Sometimes trying to find the right sound for a track can seem like a never-ending adventure, but keep at it. If nothing works, try other plugins like guitar simulators, saturation, or special effects. If nothing still seems to work, perhaps no effect was needed in the first place. Perhaps there's a more interesting track to work with?

Chapter 13
The Master Mix

While it's easy to pay attention to balancing the mix, adding EQ and effects and everything else that goes along with creating a great mix, the proper technique of the master mix buss is often overlooked. In this chapter you'll discover how to keep your mix sounding clean, how a buss compressor can make it sound a lot better or a lot worse, and how to know when your mix is finished.

Mixing With Subgroups

In an analog, digital or software console, a circuitry block that performs a function like EQ or panning is called a "stage." A big part of how the mix sounds is the way the master mix buss is driven, or what's known as "gain staging." This means that care is taken so that a stage anywhere in the mixing console isn't overloaded. When this occurs, you may hear distortion even though there's no indication of an overload on the meter or overload indicator. Let's take a look at how to correctly gain stage in order to get the cleanest sound.

As indicated in previous chapters, using subgroups either on your hardware console or your DAW makes it easy to control groups of similar sounding mix elements. Having all the drums, all the guitars, and all the background vocals each on a their own subgroup faders makes it easy to make balance changes quickly. The subgroups do call for proper gain staging for the mix signal to stay clean however, and here's how to do it.

There's a rule of thumb that works on any console, be it analog, digital, or in your DAW - *the subgroup fader levels should always be lower than the master fader level.* That means that if the master fader is set at 0dB, all the subgroup faders must be set at least -1dB or more below the master ensure that an overload doesn't occur (see Figure 13.1). If the master fader is set at -10dB for example, all the subgroup faders must be set below that point in order to avoid an overload of the mix buss (see Figure 13.2).

Figure 13.1: Correct Subgroup Gain Staging

Figure 13.2: Poor Subgroup Gain Staging

Just because your subgroups happen to be above the level of the master fader doesn't always mean that you'll get audible distortion, but the tone of the mix might be slightly altered, resulting in a smaller sounding, less punchy mix. That's why it's always a good practice to employ the "subgroup below the master" rule.

► Exercise 13.1: Mixing With Subgroups

Use the mix from the previous chapter.

A) Set the *Master Fader* to 0dB and adjust your mix so the subgroups are lower than the *Master Fader*.

☐ Does the mix sound clean?

☐ What's happening with the master mix buss meter?

B) Decrease the *Master Fader* to -20dB and raise the monitor control so that the level you're listening to is as loud as it was before.

☐ What does the mix sound like now?

☐ Has the tonal quality changed?

☐ Can you hear any distortion?

☐ What's happening with the master mix buss meter?

Individual Fader Levels

This subgroup rule can also be extended to individual faders. In order to maintain the cleanest mix with the lowest chance for an overload, always keep your individual faders below the subgroup level. If no subgroups are used, then keep them below the master fader.

The Master Level Meters

The master level meters provide an indication of whether your mix buss level is in a safe zone or the mix buss is distorting. Before we can discuss the proper meter readings a mix should have, let's look at the types of meters that are available.

Types Of Meters

There are three types of master mix buss meters that you'll find on analog and digital consoles and software consoles in a DAW. Sometimes the type of meter is selectable, and in other models or versions, it's fixed.

The VU Meter

Everyone knows what a VU meter looks like, but you see fewer and fewer of them on audio gear today (see Figure 13.3) The standard VU (which stands for Volume Units) meter, which was common on all professional audio gear until the late 90s, only shows the average level of a signal and has a response time that's way too slow for use in digital equipment. For instance, when a triangle is recorded it might read -20dB on a VU meter, yet its peaks can be as high as +10dB, which the VU meter (which is sometimes called an "RMS" meter) never indicates. Coupled with the fact that the VU is an analog, mechanical device and you see why digital metering is more useful today.

Figure 13.3: A VU meter plugin
Courtesy of TB Pro Audio

Sometimes a digital meter can be calibrated to respond like a VU meter, but unless it's designed to look like a VU meter, it's called an RMS meter instead. Regardless of it's a hardware or software meter, it's not as useful as other types of metering that are available.

The Peak Meter

The peak meter was created by the British Broadcasting Company when they realized that an engineer wasn't getting exact enough information from the VU meters it was using. This is especially important in broadcast because if a radio station allows its transmitter to go beyond its assigned transmitter wattage, it will bring a government fine. The problem was that hardware peak meters where expensive to build in the analog world, but they're quite easy in digital. As a result, most digital hardware and software consoles utilize peak meters, which respond rapidly to the signal (see Figure 13.4) and provide a true indication of what the signal is actually doing.

Figure 13.4: A Peak Meter
Courtesy of Universal Audio

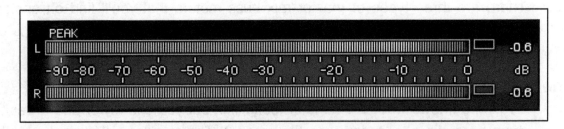

Unlike VU meters that can read beyond 0dB, digital peak meters come to a full stop at 0, which is known as "full scale" or just FS for short. Any signal beyond that point lights an overload indicator and is considered a digital "over," which may result in some very bad sounding distortion and should be avoided at all costs.

Mix Buss Levels

The Master Fader level should always be kept at a point where the meters never go beyond 0dB. Although you won't hear immediate distortion on VU/RMS or peak meters, the mix buss on some consoles and DAWs begins to change the sound of the mix from very subtle to quite noticeable as the level goes beyond that point. It's perfectly acceptable to set the Master Fader so the peaks never go beyond -10dB, and many engineers will routinely do this, especially when mixing "in the box."

By keeping the levels and the meter readings low, it assures plenty of headroom in the signal path so there's never a chance for an internal overload, which can change the sound yet never show up on the meters.

Many beginning mixers are afraid to set their mix levels low because the level of the song will be low as a result. This isn't much of a worry if your mix will be mastered, since the mastering engineer will boost the level to a commercial level.

▶ *Exercise 13.2: Setting The Mix Buss Levels*

Using the mix from the previous chapter:

A) Set the *Master Fader* so that the peaks of the mix go to about -10dB on the master buss meters.

   ☐ Does the mix sound clean?

B) Set the *Master Fader* so that the peaks of the mix go to about 0dB on the master buss meters.

   ☐ Does the mix still sound clean?

   ☐ Does it feel muddy or smaller?

C) Set the *Master Fader* so that the peaks of the mix go to about +10dB on the master buss meters (or the red Overload indicator is always lit).

   ☐ Does the mix still sound clean?

   ☐ Does it feel muddy or smaller?

D) Reset the *Master Fader* back at where the peaks of the mix go near 0dB on the master buss meters without causing the overload indicator to light.

Mix Buss Processing

Many veteran mix engineers insert a compressor on the stereo mix buss for a number of reasons:

- It raises the level of the mix and gives it more of a finished, mastered sound. Sometimes a client wants to hear what the mix will sound like after mastering, and adding some mix buss compression can simulate the effect.

- It gives the mix a sort of "glue." Many mixers will add only a dB or two of compression just to pull some of the mix elements together in a way that can't be achieved any other way.

Let's take a look at how mix buss compression is accomplished.

Mix Buss Compressor Settings

A mix buss compressor is kind of like a bit of fresh ground pepper on a nice salad; a little goes a long way (see Figure 13.5). Most mixers only add 2 or 3dB of compression on the mix and sometimes even less than that. The reason is that some compressors affect the sound of a mix in a good way without doing much compressing at all.

Figure 13.5: A UAD SSL Mix Buss Compressor plugin
Courtesy of Universal Audio

That being said, in most modern music mix buss compressors are used to make the mix "punchy" and in your face, and the trick to that is to let the attacks through without being affected while the release elongates the sound. Fast attack times reduce the punchiness of a signal, while release times that are too slow make the compressor pump out of time with the music. As mentioned in Chapter 9, *the idea is to make the compressor breathe in time with the music.*

The buss compressor is typically set at a very low compression ratio of 1.5, 2 to 1 or even 4:1, resulting in only a few dB of compression. The gain is then increased until the song's overall volume level is comparable with the hits of the genre of music you're working in.

Here are the steps to set up a buss compressor:

1) Start with the slowest attack and fastest release settings on the compressor.

2) Turn the *Attack* control faster until the high frequencies of the mix begin to dull. Stop at that point, or even back the *Attack* off a touch.

3) Adjust the *Release* control so that the compressor breathes in time with the pulse of the track.

4) Alternately, solo the snare drum and use the method for timing the compression to the track as outlined in Chapter 7.

▶ Exercise 13.3: Setting Up The Mix Buss Compressor

Using the mix you worked on in previous chapters, insert a compressor into the mix buss signal path. Set the compressor to a ratio of 2:1, the attack time as slow as it will go (0.5ms or less if calibrated in milliseconds), and the release time as fast as it will go (200ms or more if calibrated in milliseconds).

A) Adjust the *Threshold* or *Input* control until there's 2dB of gain reduction. *Make sure to adjust the Make-Up Gain so the level is the same as when the plugin is bypassed.*

☐ Does the balance of the mix change?

☐ Does the tone of the mix change?

☐ Can you hear the compression?

☐ Is there any compression indicated on the meter at all?

B) Lower the attack time until the high frequencies begin to dull, then back it off a hair. Adjust the *Threshold* control so there's still only 2dB of compression.

☐ Did anything change in the mix?

☐ Are the drums punchier

☐ Did the balance of the mix change?

☐ Can you hear the compression?

C) Raise the release time until you can hear the the compressor breathe with the track. If you can't hear it breathe, set it to half-way.

☐ Did anything change with the mix?

☐ Can you hear the compression?

☐ Can you hear any mix element better than before?

☐ Is the mix punchier?

D) Set the *Ratio* control to 4:1, but back off the *Threshold* control so that it still reads 2dB at its peak.

☐ Does the balance of the mix change?

☐ Does the tone of the mix change?

☐ Can you hear the compression?

E) Increase the *Threshold* control until there's 4dB of compression.

☐ Does the balance of the mix change?

☐ Does the tone of the mix change?

☐ Can you hear the compression?

☐ Can you hear any mix element better than before?

☐ Is the mix punchier?

F) After the compression is set, raise the *Output* control on the buss compressor until the peaks reach around -5dB on the master mix buss meter.

☐ Can you hear any difference in the mix?

☐ Does it sound better without the compressor?

G) If your mix is going to be mastered, ask the mastering engineer how he or she expects it be delivered, then mix accordingly.

► Exercise 13.4: *Setting Up the Mix Buss Compressor - Part 2*

If you're unable to set the buss compressor by listening to the entire mix, use this method instead. First, insert a compressor into the mix buss signal path. Set the compressor to a ratio of 2:1, the attack time as slow as it will go (0.5ms or less if calibrated in milliseconds), and the release time as fast as it will go (200ms or more if calibrated in milliseconds). Now try the following:

A) Solo the snare drum. Adjust the *Threshold* control until there's 2dB of gain reduction.

☐ Can you hear the compression?

☐ Is there any compression indicated on the meter at all?

B) Lower the attack time until the high frequencies of the snare begin to dull, then back it off a hair. Adjust the *Threshold* control so there's still only 2dB of compression.

☐ Did anything change in the mix?

☐ Can you hear the compression?

C) Raise the release time until the level returns to 90 to 100% by the next snare hit.

☐ Can you hear the compression?

☐ Unsolo the snare.

☐ Can you hear the compression in the mix?

☐ Can you hear any mix element better than before?

☐ Is the mix punchier?

☐ Does it sound better without the compressor?

D) Proceed with steps D through G of the previous exercise.

Mix Buss Limiting

Limiting on the mix buss is used for 2 things: prevent any overloads to the mix buss, and increase the level. When using a mix buss limiter keep the following in mind:

- Only a modern digital limiter with a Ceiling parameter control will work (see Figure 13.6) Emulations of analog gear like the UREI 1176 or Fairchild 670 don't have this.

- It must always be the last in the signal chain on the master buss.

Figure 13.6: The Waves L2 Limiter Plugin With Ceiling Control

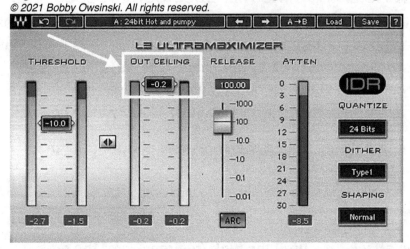

By using the combination of compressor and limiter, the level of your mix can be competitive with major label releases for purposes of comparison, although it's no substitute for professional mastering.

Exercise 13.5: *Setting Up The Limiter*

Insert a limiter at the end of the signal chain on your mix buss.

A) Adjust the *Ceiling* control to -.1dB.

☐ Can you hear any additional compression in the mix?

☐ Is the mix buss overloading?

☐ Is the mix punchier?

☐ Did the balance change?

☐ Did the sound of the midrange change?

B) Adjust the *Threshold* control until there's 6dB of gain reduction.

☐ Can you hear the compression in the mix?

☐ Can you hear any mix element better than before?

☐ Is the mix buss overloading?

☐ Is the mix punchier?

☐ Are there any dynamics left in the mix?

☐ Does it sound better or worse than with only a few dB of limiting?

☐ Does it sound better without the limiter?

C) Adjust to taste, but don't hypercompress!

Stay Away From Hypercompression!

Hypercompression is severe over-compression. Although hypercompression seems to be common today, it's very undesirable because it robs the song of any life because there are no dynamics left, and dynamics are a part of what makes music interesting. Radio stations have also proven it can cause ear fatigue, causing a listener to change to a different station in less than a minute (see Figure 13.7).

TIP: Compression should be used to control dynamics, not eliminate them.

Figure 13.7: A Dynamic Mix Versus A Hypercompressed Mix

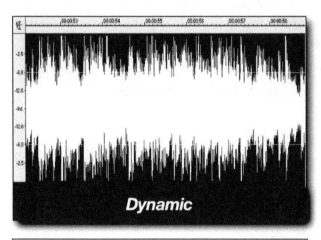

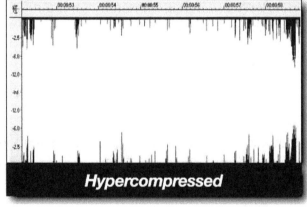

If you're going to have your mix mastered, about the worse thing that you can do is to over-compress it because there's very little for the mastering engineer to work with. Mastering engineer's are used to getting compressed mixes in these days where about 10 years ago that would never happen. Most mixers would bypass the buss compressor before sending it to mastering.

Exercise 13.6: The Evils Of Hypercompression

Begin with the same setup as in exercise E13.3 and follow steps A through D.

A) Adjust the compressor's *Threshold* control until there's 8dB of gain reduction.

- ☐ Can you hear the compression in the mix?
- ☐ Can you hear any mix element better than before?
- ☐ Is the mix buss overloading?
- ☐ What happens to the limiter?
- ☐ Is the mix punchier?
- ☐ Are there any dynamics left in the mix?
- ☐ Does it sound better or worse than with only a few dB of compression?

B) Adjust the *Threshold* control until there's 20dB of gain reduction.

- ☐ Can you hear the compression in the mix?
- ☐ Can you hear any mix element better than before?
- ☐ Is the mix punchier?
- ☐ Is the mix buss overloading?
- ☐ What happens to the limiter?
- ☐ Are there any dynamics left in the mix?
- ☐ Does it sound better or worse than with only a few dB of compression?

C) Adjust to taste, but don't hypercompress!

How Long Should My Mix Take?

While many beginning mixers will fly through a mix and be finished in an hour, most mix veterans take a lot longer than that to get a mix together. As you've seen from the previous chapters, a fair amount of experimentation is required to find the parameter settings for EQ, compression and effects that work with the track.

While rough mixes are done very fast by nature (they may only take a couple of passes through the song to get together), and you might occasionally get lucky with a quick mix, most veteran mixers usually figure it takes a five to twelve hour day to mix a song. Many mixers like to have an extra half-day to tweak things with a fresh ear, but many big-name, big-budget projects can take weeks of multiple mixes to sculpt it just right.

Of course, if all you have is time, then you can mix a song way beyond its peak (the final mix for Michael Jackson's hit "Billy Jean" was #2 out of 99), so it's best to have a few guidelines as to when it might be finished.

How To Know When Your Mix Is Finished

You can consider a mix finished when the following occur:

- **You can feel the groove.** Whatever element supplies the groove, it has to be emphasized so that the listener can feel it.

- **Every instrument or vocal can be clearly heard.** If an instrument or vocal is unintentionally masked or covered by another instrument or vocal, then your mix isn't finished.

- **Every lyric and every note of every line or solo must be clearly heard.** Each note should be crystal clear. Tweak your fader automation to help this out.

- **Be sure the mix is punchy.** This is usually a function of the bass and drum EQ and compression.

- **The mix has an interesting element.** Make sure the most important element of the song is obvious to the listener.

- **Be sure your mix sounds good when you play it against other songs that you like.** Consider it a job well done when this happens.

If time is not a problem, do as many versions as you can until you feel satisfied with your creation. Mixing takes experience so the more time you put in, the better you get at it.

Exercise 13.6: *Is My Mix Finished?*

Use the mix that you've been working on in the previous chapters of the book.

A) Listen to the mix.
- ☐ Is the groove obvious?
- ☐ Can you feel it in the mix?

B) Listen to the mix.
- ☐ Can every instrument be clearly heard?

C) Listen to the mix but concentrate on the vocal.
- ☐ Can every lyric be heard?

D) Listen to the drums.
- ☐ Are they punchy?

E) Listen to the mix.
- ☐ What grabs your attention?

F) Play your mix and then play your favorite song.
- ☐ When both songs are played back to back, does your mix stand up?

◆

242

Chapter 14
Building The Mix In Order

While most topics in the book where compiled together in a way to make the concepts easier to grasp, the order that we apply the techniques during a mix is slightly different. Now that you know how the different elements of building a mix work, let's do it in an order that makes it fast and easy.

Visualize Your Mix

Most mixers can hear some version of the final product in their heads before they get too far into the mix. This is because they've heard rough mixes of the song many times already during production. Even if a mixer is brought in just for the mix, they listen to all the elements several times before they really get down to mixing.

If you're just starting out mixing, you might think, "How can I hear the final product before I've even begin?" That's a fair enough question. Until you have a certain amount of experience, you need a few questions to help mold your vision a bit, and the way to do that is to go back to the six mix elements and ask yourself:

- Can I hear the final balance in my head?

- Can I hear how the mix elements will be EQed"

- Can I hear how everything will be panned?

- Can I hear how everything will be compressed?

- Can I hear how ambience and effects can be used in the song?

- What do I hear as the most interesting thing in the song?

If you can answer these questions, you still may not have a full picture of your final mix. You'll have at least a general idea though, which is the first step to a great mix.

TIP: Keep in mind that the producer, musicians and maybe the songwriters have a say in the mix as well, and your version of the mix can suddenly take a wide left turn with their input. That's okay, because after you've gotten everything to the point where you hear it in your head (or even beyond), a left turn should be easy.

► Exercise 14.1: *Visualizing Your Mix*

Now let's think only about the balance of the various mix elements.

A) Either listen to a rough mix of the song you're working on, or quickly just push up all the faders for a rough balance.

☐ How loud do you hear the drums or drum loops in the final mix?

☐ How about the bass element?

☐ Do you hear the vocals out in front of the other mix elements, or tucked back in the track?

☐ How loud do you hear the primary musical elements that carry the song?

☐ How loud do you hear the secondary elements like percussion and background vocals?

B) Now let's think about the frequency response of the various tracks.

☐ Is there a mix element or two that sounds particularly dull?

☐ Is there a mix element or two that sounds overly bright?

☐ Is there a mix element that has too much bottom end?

☐ Is there a mix element that doesn't have enough bottom end?

C) Now let's think about the panning.

☐ How do you hear the drums or drum loops panned? Wide or narrow?

☐ How do you hear the panning of any mix elements or loops that were recorded in stereo?

☐ Do you hear any mix elements panned extreme wide left and right?

☐ What mix elements do you hear panned up the center?

D) Now let's go to compression.

☐ Is there an instrument, loop, or vocal that has wild dynamic shifts that needs compression?

☐ Is there an instrument, loop, or vocal that you'd like to change the sound of by using compression?

☐ Is there an instrument, loop, or vocal that needs to sound a little more punchy?

E) Let's think about the ambience.

☐ What mix elements were already recorded with room ambience or reverb?

☐ Do you hear ambience on the drums, snare, or claps?

☐ What mix elements do you hear rather dry and in your face?

☐ What mix elements do you hear further away from you?

F) Lastly, it's time to think about the interest element of the song.

☐ What's the most important element in the mix?

☐ If there isn't one yet, can you create one?

☐ What's the next most important element in the mix?

☐ What's the next most important element in the mix after that?

These aren't all the questions that you can ask yourself about a mix, but you get the idea. Remember, there are no right or wrong answers. It's as you visualize it in your head.

The 10 Steps To Creating A Mix

Here are 10 steps to creating a great mix. It's best to do these in order to eliminate the time it takes unnecessarily tweaking.

Step 1: Prep Your Tracks

This includes editing out noises, deleting or hiding unused tracks, renaming tracks as needed, reordering tracks on the mix page, and anything else you need to make the tracks mix-ready. Then, insert effects channels you think you'll need and as well as groups and subgroups.

Step 2: Insert Master Buss Processing

Now insert any processors on the master buss that you think you'll be using during the mix. *Bypass everything but the compressor at first*, as that will affect the balance of the mix and we'll have to tweak it later anyway.

If you have a favorite preset for the compressor, dial it up now, otherwise use the default setting.

Step 3: Insert The High Pass Filter

Insert a high pass filter on all tracks and roll off the low frequencies as needed.

Steps 4 and 5: Set levels And Set Compression

These 2 steps actually go hand in hand and doing both at the same time makes the mix go faster. Set the levels of each track, using either the ear or meter method, but add compression to the tracks that are not steady and contain a lot of peaks. Also, group together similar tracks like background vocals, keyboards and horns for leveling and ease of processing later.

Step 6: Tweak The Master Buss

At this point you may find that the levels on master buss meters are now much hotter and are lighting the overload indicators. Either back the master fader down until they no longer light, or insert your limiter on your master buss.

Step 7: Begin EQing

Now that you have a basic mix set up, you'll notice that certain instruments are clouded and need to be brightened. You'll hear that some instruments mask others. Solo in pairs or more and EQ until you can hear them all distinctly. Remember that when you add or subtract EQ that the level will change so you'll have to rebalance.

Step 8: Add Effects

Does the mix seem dull and uninteresting? Are there mix elements that you want to make sound bigger or push back in the track? Is there an element that's uninteresting by itself and needs motion? Now is the time to begin experimenting with delay, reverb, and modulation combinations.

Step 9: Automate

Although not covered in this book since it's an advanced technique, now is the time to automate your faders and mutes as needed. This is the last step in building your mix as it's almost finalized and needs level adjustments in certain places of the song to add interest to tweak the arrangement.

Step 10: Tweak The Buss Processors

Now that you're happy with the mix, it's time to tweak the limiter in order to get more level if needed. If you find that the mix is dull or not as full as you'd like, a touch of high and low frequency EQ here can do wonders.

On Your Mixing Journey

Of course there's no one way to build a mix, as engineers have been showing us for decades. What you've read in this book is just a reference point to get you started. Take what works for you and leave the rest behind. If you don't find a technique useful, at least you'll know what it is and why it doesn't work for you so you can move beyond it.

Above all, know that mixing as all about experience. The more you do it, the more you learn and the better you get at it.

Hit mixes have been done in less than an hour and over days, weeks and months. There is no right or wrong way to do it, or right amount of time to make it happen. It's all about presenting the song in its best possible light so more people can enjoy it. If it feels good, it is good.

Happy mixing!

About Bobby Owsinski

Music-industry veteran **Bobby Owsinski** is an in-demand producer/engineer working not only with a variety of recording artists, but also on commercials, television, and motion pictures. He is an expert on surround sound music mixing and has worked on surround projects and DVD productions for superstar acts including Jimi Hendrix, the Who, Willie Nelson, Neil Young, Iron Maiden, the Ramones, and Chicago, among many others. Bobby is also one of the best-selling authors in the music recording industry, with 25 books that are

He's also a contributor to Forbes, his popular blogs have passed 7 million visits, and he's appeared on CNN and ABC News as an audio and music branding expert. Many of his books have also been turned into video courses that can be found online at lynda.com, and he continues to provide presentations, workshops and master classes at conferences and universities worldwide.

Bobby's blogs are some of the most influential and widely read in the music business. Visit his production blog at bobbyowsinskiblog.com, his music industry blog at music3point0.com, his podcast at bobbyoinnercircle.com and his website at bobbyowsinski.com. He can be found on Forbes at forbes.com/sites/bobbyowsinski.

Other Music-Related Books By Bobby Owsinski

The Mixing Engineer's Handbook *4th Edition*

The Recording Engineer's Handbook *4th Edition*

The Mastering Engineer's Handbook *4th Edition*

Social Media Promotion For Musicians *3rd Edition -The Manual For Marketing Yourself, Your Band or your Music Online*

The Music Business Advice Book

Abbey Road To Ziggy Stardust [with Ken Scott]

Music 4.1: A Survival Guide For Making Music In The Internet Age (*5th edition*)

The Drum Recording Handbook *2nd Edition* [with Dennis Moody]

How To Make Your Band Sound Great

The Studio Musician's Handbook [with Paul ILL]

The Music Producer's Handbook *2nd Edition*

The Musician's Video Handbook

Mixing And Mastering With T-Racks: The Official Guide

The Touring Musician's Handbook

The Ultimate Guitar Tone Handbook [with Rich Tozzoli]

The Studio Builder's Handbook [with Dennis Moody]

The Audio Mixing Bootcamp

Audio Recording Basic Training

Deconstructed Hits: Classic Rock Vol. 1

Deconstructed Hits: Modern Pop & Hip-Hop

Deconstructed Hits: Modern Rock & Country

The PreSonus StudioLive Mixer Official Manual

You can get more info and read excerpts from each book by visiting the excerpts section of bobbyowsinski.com.

Bobby Owsinski Lynda.com Video Courses

Audio Recording Techniques

The Audio Mixing Bootcamp

Audio Mastering Techniques

Social Media Basics for Musicians and Bands

Bookmarking Sites for Musicians and Bands

Blogging Strategies for Musicians and Bands

YouTube for Musicians and Bands

Twitter for Musicians and Bands

Mailing List Management for Musicians and Bands

Website Management for Musicians and Bands

Facebook for Musicians and Bands

Mastering For iTunes

Music Studio Setup and Acoustics

Selling Music Merchandise

Selling Your Music: CDs, Streams and Downloads

Bobby Owsinski Online Coaching Courses

The following can be found at BobbyOwsinskiCourses.com:

101 Mixing Tricks

The Music Mixing Primer

Vocal Mixing Techniques

The Music Mixing Accelerator

Music Producer Formula

Brand Your Music Crash Course

Music Prosperity Breakthrough

Editing Tricks Of The Pros

Vintage Gear Mixing Tricks

Bobby Owsinski's Social Media Connections

Music Production Blog: bobbyowsinskiblog.com

Music Industry Blog: music3point0.com

Inner Circle Podcast: bobbyoinnercircle.com

Forbes blog: forbes.com/sites/bobbyowsinski/

Facebook: facebook.com/BobbyOwsinskiBiz

YouTube: youtube.com/user/polymedia

Instagram: instagram.com/bobbyowsinski

Linkedin: linkedin.com/in/bobbyo

Twitter: @bobbyowsinski

Website: bobbyowsinski.com

◆

FREE EXERCISE TRACKS

Download the sample tracks that go along with the mixing exercises in this book at:

bobbyowsinskicourses.com/sessions

There you'll find 2 songs with a variety of mix elements to work on in both Pro Tools and Logic sessions. You can also download the raw tracks into any other DAW of your choice.

Also, to discover more about mixing that goes beyond this book, check out the *Music Mixing Primer* and *101 Mixing Tricks* programs at BobbyOwsinskiCourses.com